Advances in Robotics and Autonomous Systems

Advances in Robotics and Autonomous Systems

Guest Editors

Zhufeng Shao
Fumin Zhang

MDPI

Basel • Beijing • Wuhan • Barcelona • Belgrade • Novi Sad • Cluj • Manchester

Guest Editors
Zhufeng Shao
Department of
Mechanical Engineering
Tsinghua University
Beijing
China

Fumin Zhang
School of Electrical and
Computer Engineering
Hong Kong University of
Science and Technology
Hong Kong
China

Editorial Office
MDPI AG
Grosspeteranlage 5
4052 Basel, Switzerland

This is a reprint of the Special Issue, published open access by the journal *Applied Sciences* (ISSN 2076-3417), freely accessible at: https://www.mdpi.com/journal/applsci/special_issues/902J5C5X36.

For citation purposes, cite each article independently as indicated on the article page online and as indicated below:

Lastname, A.A.; Lastname, B.B. Article Title. *Journal Name* **Year**, *Volume Number*, Page Range.

ISBN 978-3-7258-7434-7 (Hbk)
ISBN 978-3-7258-7435-4 (PDF)
https://doi.org/10.3390/books978-3-7258-7435-4

Cover image courtesy of Zhufeng Shao

Contents

About the Editors

Zhufeng Shao

Zhufeng Shao received his Ph.D. in Mechanical Engineering from Tsinghua University, China, in 2011. He is currently an Associate Professor with the Department of Mechanical Engineering at Tsinghua University. He is a member of the National Industrial Foundation Expert Committee, a special expert of the Chinese Academy of Engineering for strategic research on industrial foundations, a senior member of the Chinese Mechanical Engineering Society, an Editorial Board Member of the journal Defence Technology, and a member of the National Technical Committee for Professional Standards. Dr. Shao's research focuses on high-performance robotic equipment and intelligent manufacturing. His research interests include cable-driven parallel robots, soft robotics, learning-augmented modeling and control, autonomous systems, and intelligent manufacturing. Dr. Shao has published 55 SCI-indexed papers and 43 EI-indexed papers, authored two academic books, and holds 61 Chinese invention patents and 23 software copyrights. His research has led to several industrial applications, including the development of an intelligent painting system for large civil aircraft, an automated coating system for non-structural surfaces of large ship segments, and a novel cable-driven high-speed parallel robot.

Fumin Zhang

Fumin Zhang is Chair Professor at the Hong Kong University of Science and Technology, serving as Director of HKUST-BDR Joint Research Institute, Director of HKUST-JD.COM Joint Research Lab, Director of HKUST-Wuxi Joint Laboratory for Robotic Embodied Dexterity and Intelligence, and Associate Director of Low Altitude Economy Research Center. He received a PhD degree in 2004 from the University of Maryland (College Park) in Electrical Engineering and held a postdoctoral position at parallel mechanisms from 2004 to 2007. His research interests include mobile sensor networks, maritime robotics, control systems, and theoretical foundations for cyber-physical systems. He received the NSF CAREER Award in September 2009 and the ONR Young Investigator Program Award in April 2010. He is currently serving as the co-chair for the IEEE RAS Technical Committee on Marine Robotics, associate editors for *IEEE Transactions on Automatic Control*, and *IEEE Transactions on Control of Networked Systems*, *IEEE Journal of Oceanic Engineering*, and *International Journal of Robotics Research*. He is an IEEE and ASME Fellow.]

applied sciences

MDPI

Editorial

Advances in Robotics and Autonomous Systems: Transitioning from Automation to Intelligent Collaboration

Zhufeng Shao [1,2,*] and Fumin Zhang [3]

1 State Key Laboratory of Tribology in Advanced Equipment, Department of Mechanical Engineering, Tsinghua University, Beijing 100084, China
2 Beijing Key Laboratory of Transformative High-End Manufacturing Equipment and Technology, Department of Mechanical Engineering, Tsinghua University, Beijing 100084, China
3 Cheng Kar-Shun Robotics Institute, Hong Kong University of Science and Technology, Hong Kong SAR, China; eefumin@ust.hk
* Correspondence: shaozf@mail.tsinghua.edu.cn

1. Introduction

Since the Third Industrial Revolution, robotics and automated systems have become deeply integrated into industrial production and daily life, fundamentally reshaping production efficiency and product quality. With the ongoing advancement of the Fourth Industrial Revolution (Industry 4.0), a new paradigm is emerging—one defined by the seamless integration of physical systems and digital intelligence. In this evolving landscape, robots are no longer isolated machines executing predefined instructions; rather, they are transitioning into intelligent, autonomous agents capable of sustained and context-aware collaboration with humans [1].

Driven by rapid progress in Artificial Intelligence (AI) and the Internet of Things (IoT), the enabling technologies of robotics are advancing at unprecedented speed. The deep convergence of mechatronics, advanced materials, computer vision, and modern control theory provides the technological foundation for robots to transcend the constraints of structured factory environments and operate reliably in complex, dynamic, and unstructured settings. Synthesizing current research trajectories and engineering practices, the evolution of robotics and autonomous systems can be conceptually organized into four tightly coupled and mutually reinforcing layers.

Layer 1: Robotic Mechanisms and Morphology. At the foundational level, research is redefining robotic embodiment beyond traditional rigid-body architectures. Emerging paradigms include rigid–flexible hybrid systems [2], high-speed, high-precision parallel and hybrid mechanisms [3,4], bio-inspired and humanoid robots [5], and deformable soft robots [6]. While expanding application scenarios, these new forms introduce significant nonlinearity, high dimensionality, and difficult-to-model hysteretic effects, challenging traditional control theories and the entire perception-to-actuation pipeline.

Layer 2: Perception and Environmental Modeling. The deep integration of machine vision and machine learning has propelled robotic perception from geometric reconstruction toward semantic cognition [7]. Beyond achieving high-precision localization and robust trajectory estimation in dynamic, texture-poor, and low-illumination environments [8], current research increasingly emphasizes semantic mapping that integrates object attributes with contextual environmental relationships. Through multi-modal sensor fusion—including vision, LiDAR, thermal sensing, and tactile feedback—robots are developing all-weather, multi-dimensional perception frameworks capable of delivering reliable and interpretable state information to higher-level decision systems [9].

Layer 3: Control, Planning, and Digital Twins. For robots containing flexible actuators and structurally complex components that resist analytical modeling, data-driven paradigms and Reinforcement Learning (RL) have demonstrated strong adaptability and generalization capabilities [10]. Advanced control theories and optimization frameworks—such as convex optimization and Mixed-Integer Programming (MIP)—are increasingly deployed to address nonlinear motion planning and obstacle avoidance in high-degree-of-freedom systems [11]. Under hardware constraints and environmental uncertainty, planning architectures are evolving from deterministic path generation toward cognitively informed, task-level reasoning [12], while distributed coordination and Swarm Intelligence expand system scalability [13].

Moreover, the integration of Large Language Models (LLMs) introduces logical reasoning and long-horizon task planning capabilities, positioning robots as intelligent coordinators that bridge perception and execution [14]. In parallel, Digital Twin and Sim2Real technologies provide a critical infrastructure linking virtual training with physical deployment [15]. Through high-fidelity simulation and domain randomization, researchers are actively mitigating the "reality gap," accelerating the translation of theoretical advances into robust engineering solutions [16].

Layer 4: Application Systems and Societal Scenarios. The ultimate validation of robotics lies in real-world deployment across industrial manufacturing [17], underwater and field operations [18], and healthcare and elderly care services [19]. At the system level, long-term reliability, safety assurance, interoperability, and social acceptance emerge as decisive indicators of technological maturity. Importantly, application-layer demands exert reverse constraints on the underlying mechanical, perceptual, and control layers, compelling continuous refinement toward greater intelligence, compliance, and trustworthiness.

Against this backdrop, this Special Issue, entitled "Advances in Robotics and Autonomous Systems", was established as a platform for disseminating recent breakthroughs across these interconnected layers. The collected contributions span soft robot structural optimization, precision calibration of parallel mechanisms, perception and localization in mobile and underwater systems, and AI-driven societal demand analysis. Together, they reflect the dynamic evolution of modern robotics toward enhanced monitoring accuracy, control robustness, and operational reliability across diverse application domains.

2. An Overview of Published Articles

This Special Issue compiles eight high-quality research papers that concretely embody the four-layer technical framework outlined above. Through innovations in mechanism design, perception modeling, intelligent control, and application-oriented analytics, these works collectively illustrate how modern robotics advances from foundational hardware to system-level societal integration.

Layer 1: Mechanical Structure Innovation and Kinematic Optimization. Mechanical innovation remains the physical foundation of robotic intelligence. Regarding soft robotics, Khan et al. drew inspiration from reptilian scale friction to design a bio-inspired Kirigami anchoring structure, overcoming low locomotion efficiency in unstructured environments [20]. Through systematic geometric and actuation parameter optimization, this study demonstrated that specific triangular Kirigami configurations significantly enhance anisotropic friction properties. Validation demonstrated a stride efficiency of 63% and

 https://doi.org/10.3390/app16041966

a crawling speed of approximately 47 cm/min, offering a compact and energy-efficient locomotion strategy for confined and complex environments.

Studying high-precision rigid mechanisms, Yu addressed the stringent calibration requirements of parallel platforms used in spacecraft on-orbit docking simulations [21]. An Improved Particle Swarm Optimization (PSO) algorithm was developed to convert complex kinematic parameter identification into a tractable high-dimensional nonlinear optimization problem. The experimental results showed that positioning error was reduced from 4.33 mm to 0.77 mm after calibration, accompanied by a significant improvement in attitude accuracy. This work highlights the increasing role of evolutionary optimization in enhancing structural precision and dynamic fidelity.

Layer 2: Precise Perception and Robust Localization. Perception serves as the cognitive gateway between the robot and its environment. To mitigate visual occlusion in complex object manipulation tasks, Hong et al. proposed a "See-Then-Grasp" strategy featuring a Two-Stage Active Reconstruction framework using a single robotic arm [22]. By autonomously optimizing viewpoints to eliminate blind spots, the system constructs high-fidelity 3D object models, significantly improving grasp success rates for previously unseen objects in unstructured settings. This work demonstrates the shift from passive perception to active, task-driven environmental reconstruction.

For indoor mobile robot localization in GPS-denied environments, Al-Hadithi and Pastor tackled the cost barrier of conventional LiDAR- or vision-based systems [23]. They designed a low-cost localization architecture based on ultrasonic beacons and differential Time-of-Flight (dToF) measurements. By refining differential ranging algorithms to suppress cumulative ultrasonic errors, the system achieved reliable navigation performance without expensive sensing infrastructure, offering a scalable pathway for large-scale deployment of logistics and service robots.

Layer 3: Reliable Control and High-Fidelity Simulation. Control and simulation technologies form the bridge between digital intelligence and physical execution. In fine manipulation, Park et al. proposed a Cartesian force-controlled Pushing Primitive to overcome the limitations of traditional pick-and-place strategies [24]. By introducing compliant force feedback, the system enabled precise in-hand adjustment without reliance on complex visual servoing. Experimental evaluations demonstrated placement errors controlled within 0.25 mm (X-axis) and 0.23 mm (Y-axis), providing a robust non-prehensile manipulation solution for precision assembly tasks.

In extreme environment simulation, Orjales et al. addressed the dynamics modeling challenges of open-frame Remotely Operated Vehicles (ROVs) by proposing a hybrid parameter identification method combining empirical measurement and Evolutionary Algorithms [25]. Using the BlueROV2 platform, they automatically optimized hydrodynamic coefficients to build a high-fidelity Digital Twin, effectively bridging the simulation-reality gap in complex underwater environments.

Similarly, Dong et al. focused on snow-removal field robots and established a systematic calibration methodology for snow particle contact parameters [26]. Using the Discrete Element Method (DEM) combined with the Hertz–Mindlin with JKR contact model and response surface analysis, they optimized contact parameters to keep the relative error of the simulated angle of repose within 0.32%, providing a reliable virtual verification benchmark for structural optimization.

Layer 4: Societal Application Analysis and System-Level Intelligence. At the societal systems level, robotics must align with demographic and ecological realities. Wen et al. applied machine learning techniques to conduct a large-scale quantitative analysis of community elderly care demands across major Chinese cities [27]. The resulting high-

https://doi.org/10.3390/app16041966

precision predictive model revealed nonlinear patterns in health conditions and lifestyle factors among elderly populations. Although centered on data analytics, the study provides essential socio-technical evidence for the functional design and deployment strategies of next-generation nursing robots and assistive systems, reinforcing the feedback loop between application demand and technological development.

3. Conclusions

Collectively, the contributions in this Special Issue delineate a coherent evolution of robotic systems toward greater intelligence, compliance, and societal integration across multiple system layers. Interpreting these works within a unified four-layer framework clarifies not only current technical achievements but also the structural trajectory and emerging bottlenecks shaping next-generation robotics.

Mechanisms. Robotic design is transitioning from a traditional emphasis on stiffness and load capacity toward paradigms centered on rigid–flexible fusion, variable stiffness, and intrinsically safe interaction. Parallel mechanisms, soft structures, and cable-driven architectures expand operational scale and adaptability, yet simultaneously introduce cross-scale coupling effects, nonlinear dynamics, and stringent demands on modeling and control fidelity.

Perception. Robotic perception is advancing from geometric reconstruction to semantic and physical understanding. The integration of multi-modal sensing and active perception strategies enables robots to capture not only spatial information but also physical constraints, contact states, and task-relevant semantics. Such interpretable environmental representations will form the cognitive substrate for higher-level reasoning and autonomous decision-making.

Control & Simulation. The convergence of physics-based modeling and data-driven learning is emerging as the dominant paradigm for complex robotic systems. High-fidelity Digital Twins, learning-augmented dynamic models, and robust Sim2Real transfer frameworks are essential to bridging the gap between virtual training and physical deployment, particularly in safety-critical and high-uncertainty environments.

Applications and Societal Integration. As robots increasingly operate in open, dynamic, and human-centered settings, system-level reliability, safety assurance, ethical governance, and social trust become defining criteria of technological maturity. These societal constraints exert reverse pressure on the underlying mechanical, perceptual, and control layers, driving continuous co-evolution toward more transparent, compliant, and trustworthy robotic systems.

We anticipate that this Special Issue will serve as a valuable reference for the systematic advancement of robotics and autonomous systems. By fostering dialogue across mechanism design, perception intelligence, control theory, and application ecosystems, it aims to accelerate the transition of robotic technologies from laboratory innovation to robust engineering deployment and meaningful societal impact.

Acknowledgments: The authors would like to express their sincere gratitude to Fengshan Huang for the valuable academic guidance and technical discussions, and to Xiang Zhou for his helpful assistance in improving the logic and linguistic quality of this manuscript.

Conflicts of Interest: The authors declare no conflicts of interest.

https://doi.org/10.3390/app16041966

References

1. Ribeiro, J.; Lima, R.; Eckhardt, T.; Paiva, S. Robotic Process Automation and Artificial Intelligence in Industry 4.0—a Literature Review. *Procedia Comput. Sci.* **2021**, *181*, 51–58. [CrossRef]
2. Shao, Z.; Zhang, D.; Caro, S. New Frontiers in Parallel Robots. *Machines* **2023**, *11*, 386. [CrossRef]
3. Liu, H.; Huang, T.; Chetwynd, D.G.; Kecskeméthy, A. Stiffness Modeling of Parallel Mechanisms at Limb and Joint/Link Levels. *IEEE Trans. Robot.* **2017**, *33*, 734–741. [CrossRef]
4. Trautwein, F.; Dietrich, D.; Pott, A.; Verl, A. Strategy for Topological Reconfiguration of Cable-Driven Parallel Robots. *J. Mech. Rob.* **2024**, *17*, 010908. [CrossRef]
5. Sheng, Q.; Zhou, Z.; Li, J.; Mi, X.; Xiang, P.; Chen, Z.; Xu, H.; Jia, S.; Wu, X.; Cui, Y.; et al. A Comprehensive Review of Humanoid Robots. *SmartBot* **2025**, *1*, e12008. [CrossRef]
6. Li, G.; Chen, X.; Zhou, F.; Liang, Y.; Xiao, Y.; Cao, X.; Zhang, Z.; Zhang, M.; Wu, B.; Yin, S.; et al. Self-Powered Soft Robot in the Mariana Trench. *Nature* **2021**, *591*, 66–71. [CrossRef]
7. Rosinol, A.; Abate, M.; Chang, Y.; Carlone, L. Kimera: An Open-Source Library for Real-Time Metric-Semantic Localization and Mapping. In Proceedings of the 2020 IEEE International Conference on Robotics and Automation (ICRA), Paris, France, 31 May–31 August 2020; pp. 1689–1696.
8. Song, S.; Lim, H.; Lee, A.J.; Myung, H. DynaVINS++: Robust Visual-Inertial State Estimator in Dynamic Environments by Adaptive Truncated Least Squares and Stable State Recovery. *IEEE Robot. Autom. Lett.* **2024**, *9*, 9127–9134. [CrossRef]
9. Shan, T.; Englot, B.; Ratti, C.; Rus, D. LVI-SAM: Tightly-Coupled Lidar-Visual-Inertial Odometry via Smoothing and Mapping. In Proceedings of the 2021 IEEE International Conference on Robotics and Automation (ICRA), Xi'an, China, 30 May–5 June 2021; pp. 5692–5698.
10. Zhang, Z.; Meng, Q.; Cui, Z.; Yao, M.; Shao, Z.; Tao, B. Machine Learning Applications in Parallel Robots: A Brief Review. *Machines* **2025**, *13*, 565. [CrossRef]
11. Villani, V.; Pini, F.; Leali, F.; Secchi, C. Survey on Human–Robot Collaboration in Industrial Settings: Safety, Intuitive Interfaces and Applications. *Mechatronics* **2018**, *55*, 248–266. [CrossRef]
12. Huang, J.; Xu, Y.; Wang, Q.; Wang, Q.; Liang, X.; Wang, F.; Zhang, Z.; Wei, W.; Zhang, B.; Huang, L.; et al. Foundation Models and Intelligent Decision-Making: Progress, Challenges, and Perspectives. *Innovation* **2025**, *6*, 100948. [CrossRef]
13. Khargharia, H.S.; Ouali, A.; Shakya, S.; Ahmad, S. Collision Avoidance in UAV Swarms: A Learning-Centric Perspective on Collaborative Intelligence. *Neurocomputing* **2026**, *663*, 132020. [CrossRef]
14. Wang, J.; Shi, E.; Hu, H.; Ma, C.; Liu, Y.; Wang, X.; Yao, Y.; Liu, X.; Ge, B.; Zhang, S. Large Language Models for Robotics: Opportunities, Challenges, and Perspectives. *J. Autom. Intell.* **2025**, *4*, 52–64. [CrossRef]
15. Zhu, W.; Guo, X.; Owaki, D.; Kutsuzawa, K.; Hayashibe, M. A Survey of Sim-to-Real Transfer Techniques Applied to Reinforcement Learning for Bioinspired Robots. *IEEE Trans. Neural Netw. Learn. Syst.* **2023**, *34*, 3444–3459. [CrossRef] [PubMed]
16. Hwangbo, J.; Lee, J.; Dosovitskiy, A.; Bellicoso, D.; Tsounis, V.; Koltun, V.; Hutter, M. Learning Agile and Dynamic Motor Skills for Legged Robots. *Sci. Rob.* **2019**, *4*, eaau5872. [CrossRef] [PubMed]
17. Yao, M.; Zhou, X.; Shao, Z.; Wang, L. A General Energy Modeling Network for Serial Industrial Robots Integrating Physical Mechanism Priors. *Robot. Comput. Integr. Manuf.* **2024**, *89*, 102761. [CrossRef]
18. Liu, K.; Ding, M.; Pan, B.; Yu, P.; Lu, D.; Chen, S.; Zhang, S.; Wang, G. A Maneuverable Underwater Vehicle for Near-Seabed Observation. *Nat. Commun.* **2024**, *15*, 10284. [CrossRef]
19. Bohr, A.; Memarzadeh, K. The Rise of Artificial Intelligence in Healthcare Applications. In *Artificial Intelligence in Healthcare*; Academic Press: Cambridge, MA, USA, 2020; pp. 25–60. [CrossRef]
20. Khan, M.N.; Huo, Y.; Shao, Z.; Yao, M.; Javaid, U. Bioinspired Kirigami Structure for Efficient Anchoring of Soft Robots via Optimization Analysis. *Appl. Sci.* **2025**, *15*, 7897. [CrossRef]
21. Yu, D. Kinematic Parameter Identification for a Parallel Robot with an Improved Particle Swarm Optimization Algorithm. *Appl. Sci.* **2024**, *14*, 6557. [CrossRef]
22. Hong, Y.; Kim, J.; Cha, G.; Kim, E.; Lee, K. See-Then-Grasp: Object Full 3D Reconstruction via Two-Stage Active Robotic Reconstruction Using Single Manipulator. *Appl. Sci.* **2025**, *15*, 272. [CrossRef]
23. Al-Hadithi, B.M.; Pastor, C. Cost-Effective Localization of Mobile Robots Using Ultrasound Beacons and Differential Time-of-Flight Measurement. *Appl. Sci.* **2024**, *14*, 7597. [CrossRef]
24. Park, J.; Kim, J.-J.; Koh, D.-Y. Experimental Evaluation of Precise Placement with Pushing Primitive Based on Cartesian Force Control. *Appl. Sci.* **2025**, *15*, 387. [CrossRef]
25. Orjales, F.; Rodríguez-Cortegoso, J.; Fernández-Pérez, E.; Romero, A.; Diaz-Casas, V. Towards a Digital Twin for Open-Frame Underwater Vehicles Using Evolutionary Algorithms. *Appl. Sci.* **2025**, *15*, 7085. [CrossRef]

 https://doi.org/10.3390/app16041966

26. Dong, J.; Zhang, F.; Huang, F.; Man, X. Calibration of Snow Particle Contact Parameters for Simulation Analysis of Membrane Structure Snow Removal Robot. *Appl. Sci.* **2026**, *16*, 610. [CrossRef]
27. Wen, F.; Liu, Z.; Zhang, B.; Zhang, Y.; Zhang, Z.; Zhang, Y. A Machine Learning-Based Study on the Demand for Community Elderly Care Services in Central Urban Areas of Major Chinese Cities. *Appl. Sci.* **2025**, *15*, 4141. [CrossRef]

https://doi.org/10.3390/app16041966

applied sciences

MDPI

Article

Calibration of Snow Particle Contact Parameters for Simulation Analysis of Membrane Structure Snow Removal Robot

Jiangtao Dong [1], Fuxiang Zhang [1,*], Fengshan Huang [1,*] and Xiaofei Man [1,2]

1 School of Mechanical Engineering, Hebei University of Science & Technology, Shijiazhuang 050091, China; dongjt@hebust.edu.cn (J.D.); manxiaofei@hebust.edu.cn (X.M.)
2 Hebei Province Key Laboratory of Intelligent Detection and Control of Aerial Vehicles, Shijiazhuang 050035, China
* Correspondence: zhangfx@hebust.edu.cn (F.Z.); huangfengshan@hebust.edu.cn (F.H.)

Abstract

To enhance the accuracy of discrete element method (DEM) simulation for the snow removal process performed by autonomous robots on membrane structures, this study calibrated the key contact parameters of snow particles used in the simulation. Through literature research, the intrinsic parameters and contact parameter ranges for snow particles and membrane structures were determined. A discrete element model of snow particles was established, and the Hertz–Mindlin with Johnson–Kendall–Robert contact model was selected to simulate the formation process of the repose angle. Using the actual repose angle of snow particles as the target, four significant factors were identified through the P-B experiment, and other factors were set at the intermediate level. Through the steepest slope climbing experiment and response surface design, second-order response equations of the four significant factors were obtained. The optimal parameter combination was calculated as follows: the surface energy of snow particles was 0.23 J/m^2; the restitution coefficient, static friction coefficient, and rolling friction coefficient of snow–snow were 0.141, 0.05, and 0.03; and the restitution coefficient, static friction coefficient, and rolling friction coefficient of snow–membrane were 0.2, 0.18, and 0.03. The simulated repose angle was 40.62°, and the relative error with the actual repose angle was 0.32%. These calibration results are reliable and can provide a reliable simulation basis and essential data support for the optimal design of a snow removal robot and the dynamic simulation of the operation process.

Keywords: snow removal robot; parameter calibration of snow particles; discrete element method; repose angle measurement; data cleaning; Q-Q plot

1. Introduction

As a new type of green building form, membrane structures play an increasingly important role in modern life with their advantages of a large span, low energy consumption, high light transmittance, and excellent space utilization. They are widely used in greenhouses, granaries, stadiums, and other fields [1,2]. However, heavy snow load in winter can easily cause membrane structure deformation or even overall collapse, so there is a serious risk of safety accidents. At present, the coping approach still depends on artificial snow removal, which is not only inefficient and costly, but also has a very high safety risk when people work on the top of smooth and steep membrane structures. Therefore, the development of a snow removal robot that can replace manual work and handle membrane structure applications has become an urgent need to promote the intelligent upgrading of the construction field. However, due to the unique structural properties, such as the large curvature and smooth surface of a gas film building, it is very risky, costly, and difficult to conduct robot prototype tests directly on real membrane structures.

Therefore, simulation technology based on the discrete element method (DEM) has become one of the core means of developing, testing, and optimizing snow removal robots. By constructing a virtual simulation environment for robot–snow–membrane interaction instead of physical prototype testing, a variety of robot design schemes and operating parameters can be tested and optimized safely, economically, and efficiently [3]. This would greatly reduce the risk and cycle of research and development. But at present, the research in the field of DEM concerning snow particles mainly focuses on their microstructure and mechanical properties [4–6], while studies on the discrete element contact parameters of snow particles are still lacking. Since the DEM requires the selection of appropriate input parameter values for simulation analysis in order to make accurate predictions [7], and the contact parameters of snow particles are difficult to measure, it is of great significance to conduct research on the parameter calibration of snow particles. The repose angle is an important parameter for measuring the fluidity of granular materials, and it is often used as the core response variable for characterizing the flow characteristics of granular particles [8].

Given the submillimeter scale of natural snow particles, in the simulation process, it is usually necessary to simplify and enlarge the shape and size of such particles to accelerate the simulation speed [9]. Previous studies have already achieved the parameter calibration of discrete element simulation for micro-particles such as wheat flour [10] and quicklime powder [11] based on particle scaling methodologies.

This study conducted a simulation experiment on the repose angle using the discrete element method. To address the errors in traditional repose angle measurement methods, an improved repose angle measurement method was proposed, accurately obtaining the repose angle of the snow particle accumulation body. Through Plackett–Burman experiments, steepest slope climbing experiments, and the response surface method, significant analysis and optimization of the contact parameters of the magnified particles were carried out. To improve the accuracy of the results of the response surface method, a response surface model construction and optimization method was proposed. Data values that may have involved systematic measurement errors from the repose angle results were screened out by this method. Finally, the optimal discrete element simulation contact parameter combination of snow particles was obtained, providing reliable data support for the subsequent design and optimization of snow removal robots.

2. Materials and Methods

This section describes the materials, physical assumptions, and numerical methods employed in the discrete element simulation. First, the intrinsic physical properties of snow particles are introduced. Then, the discrete element modeling framework is presented, including the contact model selection, particle scaling theory, and simulation parameter settings. Finally, the experimental design strategies combining Plackett–Burman screening, steepest ascent, and response surface methodology are detailed, forming the basis for subsequent parameter identification and optimization.

2.1. Intrinsic Parameters of Snow Particles

The microscopic properties of granular materials encompass particle size distribution, morphological characteristics, intrinsic density, elastic modulus, and Poisson's ratio, while

https://doi.org/10.3390/app16020610

macroscopic parameters include bulk density, repose angle, and flowability indices, among others [12]. Snow is essentially an aggregate of countless tiny ice particles, and the characteristics of ice particles directly determine the macroscopic physical properties of snow [13]. According to snow scientists [14–18], snow can be roughly divided into several categories according to different factors such as temperature, time of formation, and porosity, among other factors, as shown in Table 1.

Table 1. Common snow types and their physical properties.

Type	Temperature (°C)	ρ (kg/m^3)	Porosity (%)	Cohesive Force
Fresh powdery snow	<−10	50~100	>70	Low
Round grained snow	−15~5	100~200	40~60	Low
Lightly compacted snow	−10~0	200~400	30~40	Medium
Wet snow	0~+5	350~500	30~40	High
Ice snow	−5~+5	400~600	<20	High
Firn	−10~+5	500~700	<20	High

The purpose of designing snow removal robots is to start snow removal work in a timely manner when it first snows. Since the moisture content of the first snow is relatively low, the intrinsic parameters of "dry snow" particles were investigated. During the falling process, "dry snow" undergoes equilibrium metamorphism to form circular particles [18]. Although natural dry snow particles exhibit multi-scale particle size distribution, usually from 100 μm to 1000 μm [19], the particle scaling theory is often used to save DEM calculation time for similar microparticles. However, after scaling, the smaller particles in the distribution are still relatively small, while the larger particles, although the individual effect is more significant, account for a relatively small proportion in the overall particle group. Therefore, the single dispersion system method based on the average size is often selected as a simplification method in studies [10,11]. Therefore, in this study, the mean diameter is taken as 550 μm.

In North and Northwest China, the snow density typically ranges from approximately 140 to 150 kg/m^3, as the climate in these regions is generally dry and snow cover does not persist for long periods [20]. This is consistent with the characteristics of dry snow. In the present study, an average value of 145 kg/m^3 is adopted as the macroscopic density of snow. Snow is a granular material composed of ice particles, and the density of ice is 920 kg/m^3. However, actual snow is composed of ice particles and voids, and has about 45% porosity. In the simulation, the density assigned to snow particles represents an equivalent homogeneous snow particle density, whose scale lies between the macroscopic snow sample and the microscopic individual snow grain. This equivalent density is determined using a volume-fraction-weighted averaging approach [5], given by

$$\rho_{eq} = \rho_1 \times \varphi + \rho_2 \times (1 - \varphi) \tag{1}$$

where ρ_{eq} is the equivalent particle density used in the discrete element simulations, ρ_1 is the macroscopic snow density, ρ_2 is the ice particle density, and φ is the porosity of snow. Based on this formulation, the calculated value of ρ_{eq} is 571.25 kg/m^3.

According to previous experimental studies, the Poisson's ratio of ice is approximately 0.3 and the Young's modulus is about 100 MPa [21]. Eidevåg et al. [19] tested 455 μm snow particles at −10 °C, obtaining repose angles of 40 ± 3°, 40 ± 3°, 42 ± 2°, and 41 ± 2°, averaging 40.75 ± 2.5°, which is used here as the actual repose angle.

https://doi.org/10.3390/app16020610

2.2. Discrete Element Accumulation Body Simulation Experiment

2.2.1. Simulation Experimental Device Model

Based on Eidevåg et al. [19], a 3D model was built in SolidWorks 2020. The lower snow collection platform had a diameter D = 50 mm, with the upper snow outlet located H = 100 mm above, as shown in Figure 1.

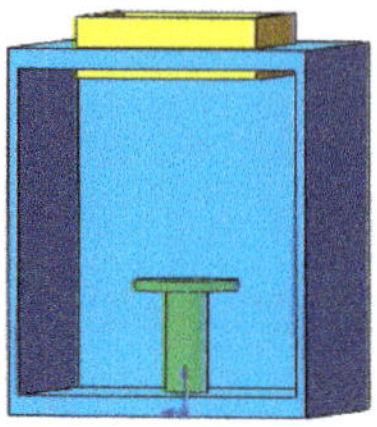

Figure 1. Simulation experimental device model.

2.2.2. Contact Model Selection

The contact model is central to the DEM. Johnson et al. [22] proposed a Hertz–Mindlin with JKR contact model for adhesive particles in 1971, and the contact model is shown in Figure 2. Although the snow investigated in this study is classified as "dry snow" under −10 °C conditions, this classification refers to a low macroscopic liquid water content rather than the complete absence of liquid water, which gives rise to surface-energy-driven adhesive interactions during particle contact [23]. On this basis, the Hertz–Mindlin contact model with JKR adhesion is adopted in this study.

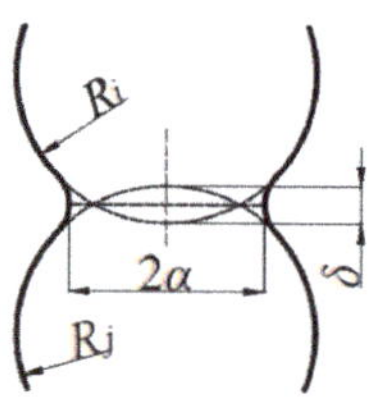

Figure 2. Hertz–Mindlin with JKR contact model.

Where R_i and R_j are the radius of the contact particles, m; α is the contact radius, m; and δ is the normal deformation, m. Due to the distribution of contact stress, the calculation formula for the total contact normal force is:

$$F_J = \frac{4E^*\alpha^3}{3R^*} - \sqrt{8\pi\gamma^* E^* \alpha^3} \tag{2}$$

where R^* is the effective contact radius and the second term is the adhesion component; E^* is equivalent Young's modulus, Pa; γ^* is the effective surface energy per unit area, J/m^2; and α is the equivalent radius, m.

$$E^* = \left(\frac{1-\nu_1}{2G_1} + \frac{1-\nu_2}{2G_2}\right)^{-1} \tag{3}$$

https://doi.org/10.3390/app16020610

where G_1 and G_2 are the shear modulus (MPa) of the two particles and ν_1 and ν_2 are the Poisson's ratio of the two particles.

$$R^* = \left(\frac{1}{R_i} + \frac{1}{R_j}\right)^{-1} \tag{4}$$

For different material particles:

$$\gamma^* = \gamma_1 + \gamma_2 - 2\gamma_{12} \tag{5}$$

where γ_1 and γ_2 are the surface energies of the two particles, J/m^2, and γ_{12} is the interfacial energy, J/m^2.

When the two granular materials are the same, $\gamma_1 = \gamma_2 = \gamma$, $\gamma_{12} = 0$, that is, $\gamma^* = 2\gamma$. According to JKR theory, the normal deformation δ is calculated as follows:

$$\delta = \frac{a^2}{R^*} - \sqrt{\frac{4\pi\gamma\alpha}{E^*}} \tag{6}$$

2.2.3. Particle Scaling Theory

In DEM, particle size can be enlarged to save computation time [24]. Dimensional analysis is employed to ensure physical consistency before and after scaling. Density is a directly measured fundamental parameter. The base dimensions are length (L), density (ρ), and time (T). Let the scaling factor for length and time be h, with density unchanged, that is [25]:

$$\lambda_L = h,\ \lambda_T = h,\ \lambda_\rho = 1 \tag{7}$$

Based on the above settings, the scale factors of other relevant particle characteristic parameters can be derived by dimensional analysis.

For the mass, M:
$M = \rho L^3$, then: $\lambda_M = \lambda_L^3 = h^3$
According to Newton's second law:
$F = Ma$, then: $\lambda_F = \lambda_\rho \lambda_L^4 \lambda_T^{-2} = h^2$
From the velocity formula:
$v = L/T$, then: $\lambda_v = \lambda_L \lambda_T^{-1} = 1$
Young's modulus:$E = \frac{FL}{Au}$, then: $\lambda_E = \lambda_\rho \lambda_L^2 \lambda_T^{-2} = 1$
Particle stress $\sigma = F/A$, then: $\lambda_\sigma = \lambda_\rho \lambda_L^2 \lambda_T^{-2} = 1$
Particle strain $\varepsilon = \sigma/E$, then: $\lambda_\varepsilon = \frac{\lambda_\sigma}{\lambda_E} = 1$
By transforming Equation (2) into a stress–strain formulation,

$$\sigma(\varepsilon, R^*) = \frac{F_J}{A} = \frac{4E^*}{3}\varepsilon^{\frac{3}{2}} - \sqrt{\frac{8\pi\Delta\gamma E^*}{R^*}}\varepsilon^{\frac{3}{4}} \tag{8}$$

It can be observed that the first term on the right-hand side is scale invariant with respect to particle size. In contrast, the second term originates from the JKR adhesive contact model and explicitly depends on the effective particle radius R^*. As a result, the JKR adhesive contribution does not satisfy scale invariance.

Based on the above similarity theory and dimensional analysis, when the particle scaling approach is employed in discrete element simulations, the scaling factors of Young's modulus, stress, strain, and velocity are equal to unity. This implies that the intrinsic material properties can be preserved after scaling. However, due to the explicit radius dependence of the JKR adhesive term, the parameters associated with the contact model

 https://doi.org/10.3390/app16020610

still require calibration to ensure that the discrete element model accurately reproduces the macroscopic flow behavior of snow. In the present study, the particle diameter is scaled from d = 0.55 mm to d = 3.3 mm, corresponding to a geometric scaling factor of six, and the simulation apparatus is scaled accordingly.

2.2.4. Simulation Parameters Setting

DEM parameters include the intrinsic properties of particles and device, contact parameters, and computational settings. These parameters significantly affect both the efficiency and accuracy of the simulation [26]. Inflatable membrane structures typically use PVC film, with ν = 0.45, E = 113 MPa [27], and ρ = 1300 kg/m^3 [28].

In DEM, a large time-step may lead to numerical inaccuracies or even instability, while an excessively small time step greatly increases computation time. The timestep is usually 10% to 20% of the Rayleigh timestep, which is calculated as [29]:

$$t_{Ray} = \frac{\pi R}{0.8766 + 0.163\nu}\sqrt{\frac{\rho}{G}} \tag{9}$$

where ν is the Poisson's ratio of the particle material; G is the shear modulus; ρ is the density of the granular material; and R is the radius of the particle.

Lommen et al. [30] found that a G varying between 10 MPa and 100,000 MPa has minimal impact on the repose angle. As Equation (7) shows, a smaller G allows faster simulation. Therefore, to maximize computational efficiency while maintaining accuracy, a shear modulus of G = 10 MPa is adopted for snow particles in this study.

2.3. Simulation Experiment Design

Experimental design is a widely employed methodology in various scientific disciplines for identifying key factors and developing predictive models. In this study, the following three-step approach is employed to identify optimal parameters: Plackett–Burman (P-B) test, steepest ascent, and response surface methodology (RSM). All designs were generated with Design-Expert 11.

Initially, a P-B test is used to distinguish significant factors among seven parameters (X0 to X6), including surface energy, contact parameters, and so on, using repose angle as the response. These parameters characterize the dynamic physical properties of particles in the simulation, and they need to be determined for the discrete element simulation software but cannot be measured directly [31].

Among these parameters, the restitution coefficients are velocity-dependent. Vázquez et al. [32] reported that the terminal velocity of snow particles, with sizes ranging from 0.06 to 3.2 mm, falls between 0.06 and 1.6 m/s. Because the wind speed at the top of the air-supported membrane structure is typically higher, the range for the coefficient of restitution was determined based on a reference terminal velocity of 1.6 m/s. Based on prior research on snow [5,6,21], the ranges of contact parameters are listed in Table 2. In the table, the minimum, maximum, and middle values of the contact parameter ranges are coded as −1, 1, and 0, respectively.

https://doi.org/10.3390/app16020610

Table 2. Plackett–Burman test parameters and coded levels.

No.	Parameters	Codes		
		−1	0	1
X0	Snow particle surface energy	0.080	0.165	0.250
X1	Snow–snow restitution coefficient	0.050	0.125	0.200
X2	Snow–snow static friction coefficient	0.050	0.125	0.200
X3	Snow–snow rolling friction coefficient	0.010	0.030	0.050
X4	Snow–membrane restitution coefficient	0.100	0.200	0.300
X5	Snow–membrane static friction coefficient	0.100	0.200	0.300
X6	Snow–membrane rolling friction coefficient	0.010	0.030	0.050

Then, the steepest ascent direction is determined based on the signs of significant effects, and step sizes are determined from parameter ranges.

Finally, a Central Composite Design (CCD) is used to establish a second-order regression model between significant parameters and the response, yielding the optimal parameter setting that is closest to the real repose angle [33].

3. Results and Analysis

This section presents the simulation results and corresponding analyses. An improved measurement method of the repose angle is first introduced and validated. Subsequently, the results of the Plackett–Burman screening experiment and the steepest ascent test are analyzed to identify significant factors and appropriate parameter ranges. The response surface modeling results, including regression analysis, outlier correction, and ANOVA, are then discussed in detail. Finally, the optimal parameter combination is determined and its physical rationality is examined through comparison with the existing literature.

3.1. Improved Measurement Method of Repose Angle

In the actual experimental determination process of the repose angle, the irregular shape formed by the accumulation body of particles often results in the measured repose angle being lower than the true value [8]. The approximate shape of the accumulation body is shown in Figure 3. This shape is a typical contour formed when measuring the repose angle of granular materials by standard experimental methods such as the Funnel Method, which provides controllable and repeatable geometric conditions for parameter calibration [7–11].

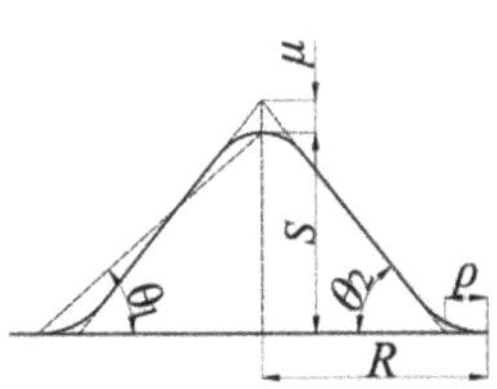

Figure 3. Actual accumulation body model.

Here, θ_1 is the measured repose angle; θ_2 is the true repose angle; S is the actual height of the accumulation body; R is the radius of the accumulation body; μ is the height error; and ρ is the radius error.

The expression of the repose angle θ_1 measured by the traditional method is:

$$\tan^{-1}\theta_1 = \frac{S}{R} \tag{10}$$

https://doi.org/10.3390/app16020610

The expression of the true repose angle θ_2 is:

$$\tan^{-1}\theta_2 = \frac{S+\mu}{R-\rho} \tag{11}$$

Due to the existence of height error μ and radius error ρ, the measured value of the repose angle will be smaller than the real value.

To address this issue, a method for measuring the repose angle of accumulation body images is proposed. In the actual accumulation contour, the slope representing the true repose angle is nearly constant, meaning the second derivative within this interval approaches zero. Therefore, according to this law, the interval segment that can express the true repose angle in the polynomial fitting function of the accumulation body contour can be found.

Compared with the method of directly fitting the half-edge contour of the accumulation body with a straight line, the proposed method can not only describe the actual shape of the accumulation body more accurately, but also avoid the interference caused by the irregular contour to a great extent. As shown in Figure 4, the specific operation steps are as follows:

1. Read the target image and convert it to grayscale, then apply binarization;
2. Detect the pile contour in the binarized image using the Canny operator;
3. Crop the contour to retain only the upper slope region and fit it with a polynomial, as shown in Formula (12);

$$y = a_n x^n + a_{n-1}x^{n-1} + \cdots + a_1 x + a_0 \tag{12}$$

   where x and y are the horizontal and vertical coordinates of the contour curve.
4. Calculate the R^2 to evaluate the fit and determine the final fitting function;

$$R^2 = 1 - \frac{\sum\limits_{i=1}^{n}(y_i-\hat{y}_i)^2}{\sum\limits_{i=1}^{n}(y_i-\overline{y})^2} \tag{13}$$

   where y_i are the actual values; $\hat{y}_i$ are the values of fitting curve at x_i; $\bar{y}$ are the mean of the actual value; and n is the total number of the data.
5. Calculate the second derivative of y, and consider the continuous intervals where the $|y''| < \varepsilon$ to be linear intervals;
   where ε reflects the straightness of the interval. After numerous experimental verifications, the test results show that $\varepsilon = 0.001$ yields satisfactorily accurate measurement results.
6. In order to ensure that the intervals have physical significance rather than local fluctuations, set the interval where the continuous horizontal distance in the linear interval exceeds 30% of the total width at the bottom of the accumulation body as the effective linear interval. Multiple simulation and experimental results demonstrate that this coverage adequately captures the overall stable geometric characteristics of the pile, and the resulting angle of repose exhibits good robustness against asymmetry and geometric noise.
7. In the effective linear intervals, obtain the average value $\bar{k}$ of the first derivative y' of all points in the 'effective linear interval'. Then calculate the repose angle θ using Formula (14).

$$\theta = \left|\tan^{-1}\bar{k}\right| \tag{14}$$

https://doi.org/10.3390/app16020610

3.2. Results of the P-B Experiment

Twelve orthogonal tests were designed by Design-Expert 11 and simulated in EDEM 2023.1. The repose angle of each test was measured, as shown in Table 3. The results of the significance analysis are presented in Table 4. In the table, the contribution (%) represents the ratio of the sum of squares of each parameter to the total sum of squares. The rank indicates the relative importance of each parameter, with Rank 1 corresponding to the largest contribution.

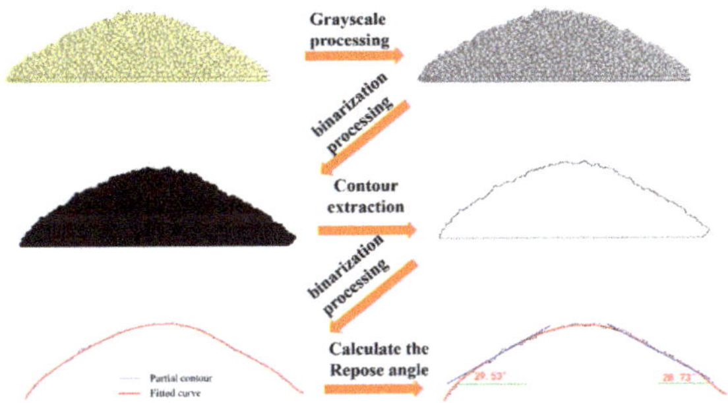

Figure 4. Measurement process of repose angle.

Table 3. Designs and results of Plackett–Burman screening experiment.

No.	Parameters							θ/°
	X0	X1	X2	X3	X4	X5	X6	
1	−1	−1	−1	1	−1	1	1	31.49
2	−1	−1	1	−1	1	1	−1	30.45
3	−1	1	−1	1	1	−1	1	36.68
4	−1	1	1	1	−1	−1	−1	33.49
5	1	1	−1	1	1	1	−1	43.80
6	1	1	1	−1	−1	−1	1	40.36
7	1	1	−1	−1	−1	1	−1	43.68
8	1	−1	−1	−1	1	−1	1	42.36
9	1	−1	1	1	−1	1	1	36.89
10	−1	1	1	−1	1	1	1	30.63
11	−1	−1	−1	−1	−1	−1	−1	37.29
12	1	−1	1	1	1	−1	−1	37.26

Table 4. Significance analysis of contact parameters based on the contribution to repose angle.

Parameters	Effect	Sum of Square	Contribution%	Ranks
X0	7.39	163.69	63.61	1
X1	2.15	13.87	5.39	3
X2	−4.37	57.29	22.27	2
X3	−0.86	2.22	0.86	6
X4	−0.34	0.34	0.13	7
X5	−1.75	9.19	3.57	4
X6	−1.26	4.76	1.85	5
$R^2 = 0.9769$	$R^2_{\text{Adj}} = 0.9365$			

3.3. Results of the Steep Climbing Experiment

Significant factors were selected to design the steep climbing experiment. The climbing step size was determined based on the parameter range and the direction of effects,

https://doi.org/10.3390/app16020610

with positive effects leading to an increase and negative effects to a decrease. The other parameters were set at the central levels, namely X3 = 0.03, X4 = 0.2, and X6 = 0.03. The experimental scheme and simulation results are shown in Table 5.

Table 5. The steepest climb test scheme and repose angle results.

No.	Parameters				$\theta/^\circ$
	X0	X1	X2	X5	
1	0.08	0.05	0.20	0.30	21.47
2	0.11	0.08	0.17	0.26	24.92
3	0.14	0.11	0.14	0.22	27.65
4	0.17	0.14	0.11	0.18	31.10
5	0.20	0.17	0.08	0.14	36.08
6	0.23	0.20	0.05	0.10	26.68

Among these experiments, the result of No. 5 is the closest to the target angle value (40.75°), so the data of No. 5 are selected as the central level, while No. 4 and No. 6 are, respectively, set as the low-level and high-level values.

3.4. Response Surface Design Results

CCD response surface design was implemented using Design-Expert 11 software. Table 6 presents the experimental factor codes. For a CCD involving k factors, the experimental design consists of three components [33]:

$$N = 2^k + 2k + n_0 \tag{15}$$

Table 6. Coded factor levels and actual parameter values.

Codes	Parameters			
	X0	X1	X2	X5
2	0.26	0.23	0.14	0.22
1	0.23	0.20	0.11	0.18
0	0.20	0.17	0.08	0.14
−1	0.17	0.14	0.05	0.10
−2	0.14	0.11	0.02	0.06

Here, 2^k represents the number of factorial points, $2k$ represents the number of axial points, and n_0 represents the number of center points.

For $k = 4$, the design includes 16 factorial points and 8 axial points to estimate the linear, interaction, and quadratic effects of the response surface. To account for inherent variability in DEM simulations caused by random particle initialization and contact sequencing, 14 replicated center points were included to improve the estimation of pure error and enhance the reliability of the ANOVA. Consequently, a total of 38 experiments were conducted, which is sufficient for constructing and analyzing a four-factor quadratic response surface model while maintaining a reasonable balance between model accuracy and computational cost.

Based on the significant parameters identified in the P-B experiment and the optimal parameter ranges selected in the steepest ascent test, the experimental plan and results are shown in Table 7.

https://doi.org/10.3390/app16020610

Table 7. CCD test scheme and measured repose angles for four iterative rounds.

No.	Parameters				$\theta_1/(°)$	$\theta_2/(°)$	$\theta_3/(°)$	$\theta_4/(°)$
	X0	X1	X2	X5				
1	2	0	0	0	40.98	40.98	40.98	40.98
2	2	0	0	0	39.47	39.47	39.47	39.47
3	1	−1	−1	−1	35.40	35.40	35.40	35.40
4	1	1	−1	−1	35.01	35.01	35.01	35.01
5	1	1	1	−1	35.02	36.12	36.12	36.12
6	1	−1	1	1	35.70	35.70	35.70	35.70
7	1	1	−1	1	36.88	36.88	36.88	36.88
8	1	1	1	1	31.88	31.88	31.88	31.88
9	1	−1	1	−1	40.99	40.99	40.99	40.99
10	1	−1	−1	1	41.13	41.13	41.13	41.13
11	0	−2	0	0	38.53	38.53	38.53	38.53
12	0	−2	0	0	36.92	36.92	37.25	37.25
13	0	2	0	0	33.25	33.25	33.25	33.25
14	0	2	0	0	32.82	32.82	32.82	32.82
15	0	0	−2	0	37.43	37.43	37.43	37.43
16	0	0	−2	0	36.34	36.34	36.34	36.34
17	0	0	2	0	36.43	36.43	36.43	36.43
18	0	0	2	0	38.21	38.21	38.21	38.21
19	0	0	0	−2	33.09	33.09	33.09	33.09
20	0	0	0	−2	34.20	34.20	34.20	34.20
21	0	0	0	2	33.24	33.24	33.24	33.24
22	0	0	0	2	34.49	34.49	34.49	34.49
23	0	0	0	0	26.43	25.63	25.63	23.78
24	0	0	0	0	26.37	26.37	25.40	25.40
25	0	0	0	0	20.13	21.23	21.23	21.23
26	0	0	0	0	20.11	20.56	23.46	23.46
27	0	0	0	0	24.31	24.31	24.31	24.31
28	0	0	0	0	26.23	26.23	26.23	22.63
29	−1	1	1	−1	36.17	36.17	36.17	36.17
30	−1	1	−1	1	35.04	35.04	35.04	35.04
31	−1	−1	−1	1	39.78	39.78	39.78	36.17
32	−1	1	−1	−1	31.70	31.70	31.70	31.70
33	−1	−1	1	−1	34.42	34.42	34.42	34.42
34	−1	−1	1	1	35.69	35.69	35.69	35.69
35	−1	1	1	1	30.94	30.94	30.94	30.94
36	−1	−1	−1	−1	32.47	32.47	32.47	32.47
37	−2	0	0	0	31.65	31.65	31.65	31.65
38	−2	0	0	0	30.89	30.89	30.89	30.89

3.4.1. Response Surface Model Construction and the Optimization Method

When designing the response surface, it is usually assumed that y is the quality characteristic of the product or process, that is, the response, and the input variables x_1, x_2, ..., x_k have the following relationship:

$$y = f(x_1, x_2, \ldots, x_k) + \varepsilon \quad (16)$$

In the model, there is a random error ε. It is usually assumed that the random errors are independently and normally distributed, i.e., $\varepsilon \sim N(0, \sigma^2)$, and the response variable has no measurement error [34]. In practice, measurement errors are inevitable. To ensure the reliability of the prediction and optimization results of the response surface method model, we can optimize the model through residual analysis based on this feature.

First, we calculate the residual ε_i between the predicted value of the regression model and the real data. Then, we construct the reference line function of ε_i through the Q-Q

https://doi.org/10.3390/app16020610

plot [35]; then, outliers are identified through residual analysis. The specific operation steps are as follows:

1. Sort the ε_i set from small to large to obtain the sample data set $u = \{u_1, u_2, \ldots, u_i, \ldots, u_n\}$, $i = 1, 2, \ldots, n$, and calculate the theoretical quantile $Z_{(i)}$ of the data. The formula is [36]:
$$Z_{(i)} = \Phi^{-1}\left(\frac{i - 0.375}{n + 0.25}\right) \quad (17)$$
where $\Phi(x)$ represents the cumulative distribution function of the standard normal distribution; n is the number of sample data; and i is the rank after sorting.
2. Construct the reference line function. The formula is:
$$y_{(i)} = \sigma Z_{(i)} + \mu \quad (18)$$
where $y_{(i)}$ represents the theoretical value of the rank i which follows the normal distribution $N(\mu, \sigma^2)$; σ is the standard deviation of the sample data. $\sigma = \sqrt{\frac{1}{N}\sum_{i=1}^{N}(u_i - \mu)^2}$; μ represents the mean value of the sample data. $\mu = \frac{1}{n}\sum_{i=1}^{n} u_i$.
3. Calculate the absolute values of residuals Δ_i for all sample data and their corresponding theoretical values, with the formula being $\Delta_i = |y(i) - u_i|$.
4. Calculate the probability density function of the absolute values of residuals Δ_i through Kernel Density Estimation (KDE), which is a nonparametric probability density estimation method, with the formula being [37]:
$$f(x) = \frac{1}{nh}\sum_{i=1}^{n} K\left(\frac{x - \Delta_i}{h}\right) \quad (19)$$
where x is a continuous variable representing the absolute value of residuals and $x \in [\Delta_1, \Delta_n]$; $f(x)$ is the probability density function; K is the kernel function, and a Gaussian kernel function is selected, $K(x) = \frac{1}{\sqrt{2\pi}h}e^{-x^2/(2h^2)}$; and h is the bandwidth.
5. Calculate the cumulative distribution function $F(x)$ of the absolute value of residuals. The formula is:
$$F(x) = \int_{-\infty}^{\Delta_q} f(x)d_x \quad (20)$$

Here, Δ_q represents the threshold for determining outliers.

The confidence level is set to 95%, and the threshold Δ_q is calculated by the cumulative distribution function $F(x) = 0.95$. If $\Delta_i \geq \Delta_q$, then it is considered that u_i in the original data is an outlier.

We carried out multiple parallel tests on the simulation experimental group corresponding to the residual u_i and took the average value as the correction result. However, the regression equation of Group 1 was obtained in the presence of outliers, so the screened outliers may be incorrectly screened or missed. Therefore, multiple iterations are required until no further outliers are detected.

According to the above operation process, this study carried out three iterations on the repose angle data of Group 1 of the response surface design, and finally found no outliers in the residual analysis results of Group 4. The Q-Q plots of the residuals of each group are shown in Figure 5, where the red point represents the abnormal data point, the blue represents the normal data point, and the dotted line represents the reference line of the normal distribution. It can be seen from the diagram that the residuals of

https://doi.org/10.3390/app16020610

Group 4 are almost distributed along the reference line, which strictly conforms to the normal distribution.

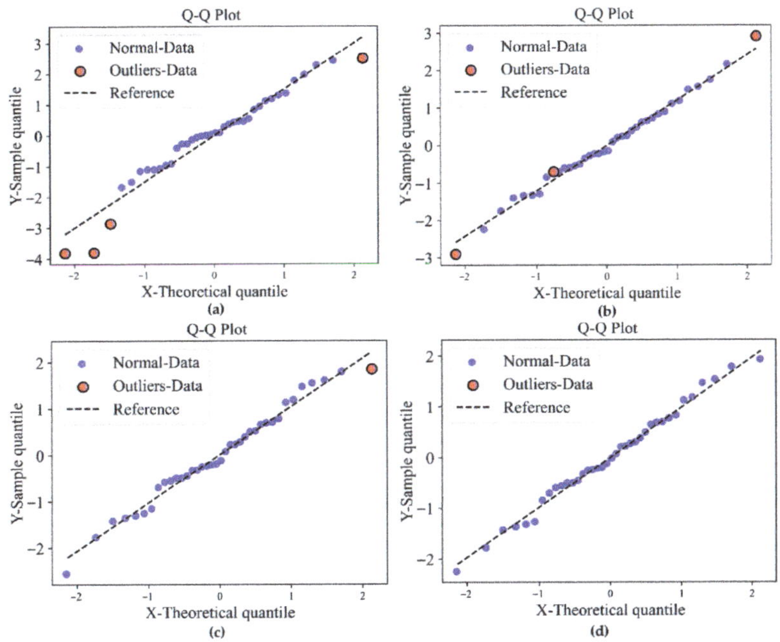

Figure 5. Result of outlier screening of Group 1 to Group 4: (**a**) Q-Q Plot of Group 1. (**b**) Q-Q Plot of Group 2. (**c**) Q-Q Plot of Group 3. (**d**) Q-Q Plot of Group 4.

In Table 7, θ_1 represents the original repose angle data of Group 1, and θ_2, θ_3, and θ_4, respectively, represent the data of Groups 2, 3, and 4 after one, two, and three iterations. The obtained regression coded equations for each group are shown in Table 8. Table 9 shows the residuals between the predicted values of the regression models from Groups 1 to 4 and the real angle of repose data.

Table 8. Coded second-order regression equations for repose angle from four iterative response surface groups.

	Group 1	Group 2	Group 3	Group 4
Constant	23.93	24.06	24.38	24.07
X0	1.610	1.650	1.650	1.650
X1	−1.300	−1.270	−1.290	−1.290
X2	−0.152	−0.1175	−0.1175	−0.1175
X5	0.211	0.1763	0.1763	0.1763
X0X1	−0.370	−0.3012	−0.3013	−0.3013
X0X2	−0.192	−0.1225	−0.1225	−0.1225
X0X5	−0.470	−0.5388	−0.5387	−0.5387
X1X2	−0.165	−0.0963	−0.0963	−0.0962
X1X5	−0.761	−0.8300	−0.8300	−0.8300
X2X5	−1.920	−1.980	−1.980	−1.980
$X0^2$	2.960	2.930	2.850	2.930
$X1^2$	2.860	2.840	2.780	2.860
$X2^2$	3.30	3.270	3.190	3.270
$X5^2$	2.460	2.560	2.470	2.550

https://doi.org/10.3390/app16020610

Table 9. Response surface model residuals of Group 1 to Group 4.

No.	Parameters				Group			
	X0	X1	X2	X5	1	2	3	4
1	2	0	0	0	1.99	1.76	1.80	1.79
2	2	0	0	0	0.48	0.25	0.29	0.28
3	1	−1	−1	−1	−1.15	−1.39	−1.41	−1.42
4	1	1	−1	−1	−0.05	−0.26	−0.24	−0.25
5	1	1	1	−1	−2.86	−1.32	−1.30	−1.31
6	1	−1	1	1	−1.49	−1.74	−1.76	−1.77
7	1	1	−1	1	0.02	−0.20	−0.18	−0.19
8	1	1	1	1	−0.12	−0.33	−0.31	−0.32
9	1	−1	1	−1	0.96	0.74	0.72	0.71
10	1	−1	−1	1	−0.26	0.11	0.09	0.08
11	0	−2	0	0	0.56	0.91	0.79	0.78
12	0	−2	0	0	−1.05	−0.70	−0.49	−0.50
13	0	2	0	0	0.48	0.27	0.23	0.22
14	0	2	0	0	0.05	−0.16	−0.20	−0.21
15	0	0	−2	0	0.00	0.40	0.40	0.39
16	0	0	−2	0	−1.09	−0.69	−0.69	−0.70
17	0	0	2	0	−0.40	−0.58	−0.58	−0.59
18	0	0	2	0	1.38	1.20	1.20	1.19
19	0	0	0	−2	−0.26	−0.48	−0.44	−0.45
20	0	0	0	−2	0.85	0.63	0.67	0.66
21	0	0	0	2	−0.95	−0.59	−0.55	−0.56
22	0	0	0	2	0.30	0.66	0.70	0.69
23	0	0	0	0	2.50	2.17	1.85	0.31
24	0	0	0	0	2.44	2.91	1.62	1.93
25	0	0	0	0	−3.80	−2.23	−2.55	−2.24
26	0	0	0	0	−3.82	−2.90	−0.32	−0.01
27	0	0	0	0	0.38	0.85	0.53	0.84
28	0	0	0	0	2.30	−0.83	−1.15	−0.84
29	−1	1	1	−1	1.33	1.12	1.14	1.13
30	−1	1	−1	1	0.10	0.49	0.51	0.50
31	−1	−1	−1	1	1.79	−0.21	−0.23	−0.24
32	−1	1	−1	−1	0.44	0.22	0.24	0.23
33	−1	−1	1	−1	−1.09	−1.33	−1.35	−1.36
34	−1	−1	1	1	1.13	1.50	1.48	1.47
35	−1	1	1	1	0.10	−0.13	−0.11	−0.12
36	−1	−1	−1	−1	1.21	1.57	1.55	1.54
37	−2	0	0	0	−0.90	−0.53	−0.49	−0.50
38	−2	0	0	0	−1.66	−1.29	−1.25	−1.26

3.4.2. ANOVA for Corrected Regression Model

The analysis of variance (ANOVA) was performed on Group 4, and the second-order regression model, as a whole, was very significant ($p < 0.0001$, $R^2 = 0.9529$, $R^2_{\text{Adj}} = 0.9242$). The lack-of-fit term ($p = 0.2445 > 0.05$) confirmed the model's adequacy and absence of systematic bias, with detailed results presented in Table 10. The response surfaces of the interaction between X0 and X1, X2, and X5 are shown in Figure 6.

https://doi.org/10.3390/app16020610

Table 10. ANOVA for regression model analysis of variance.

Source	Sum of Squares	Degrees of Freedom	Mean Square	F	*p*
Model	910.71	14	65.05	33.24	<0.0001 **
X0	86.86	1	86.86	44.38	<0.0001 **
X1	53.20	1	53.20	27.19	<0.0001 **
X2	0.4418	1	0.4418	0.2258	0.6392
X5	0.9941	1	0.9941	0.5080	0.4832
X0X1	1.45	1	1.45	0.7420	0.3979
X0X2	0.2401	1	0.2401	0.1227	0.7293
X0X5	4.64	1	4.64	2.37	0.1371
X1X2	0.1482	1	0.1482	0.0757	0.7856
X1X5	11.02	1	11.02	5.63	0.0264 *
X2X5	62.96	1	62.96	32.17	0.00012 **
$X0^2$	355.90	1	355.90	181.87	<0.0001 **
$X1^2$	338.79	1	338.79	173.12	<0.0001 **
$X2^2$	443.02	1	443.02	226.38	<0.0001 **
$X5^2$	245.11	1	245.11	125.25	<0.0001 **
Residual	45.01	23	1.96		
Lack of Fit	24.08	10	2.41	1.50	0.2445
Pure Error	20.93	13	1.61		
Total	955.72	37			

Note: * means significant; ** means very significant.

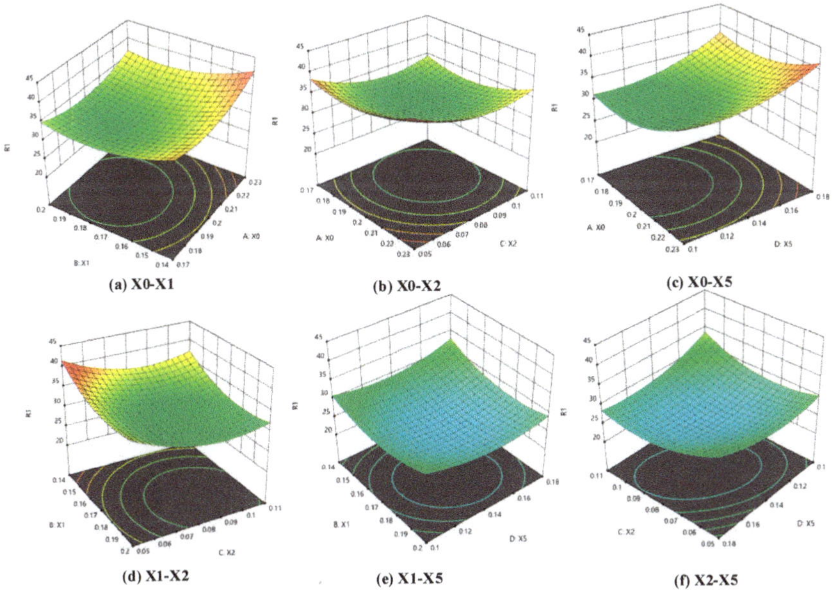

Figure 6. Effects of factor interactions on the repose angle: (**a**) Interaction between X0 and X1. (**b**) Interaction between X0 and X2. (**c**) Interaction between X0 and X5. (**d**) Interaction between X1 and X2. (**e**) Interaction between X1 and X2. (**f**) Interaction between X2 and X5.

In the ANOVA, the terms X2 and X5 are not significant in their linear contributions but are highly significant in the corresponding quadratic terms. This result indicates that the response variable does not vary with the parameters in a simple proportional manner. From a microscopic physical perspective, the macroscopic repose angle of snow particles is

https://doi.org/10.3390/app16020610

governed by the structural stability of the particle contact network and the associated energy dissipation mechanisms, which are coupled during particle deposition and rearrangement processes and collectively determine the stable angle of the accumulation body. When the relevant parameters fall within certain ranges, the particle system may undergo a gradual transition from a friction-dominated contact regime to an adhesion–friction coupled regime. Such a transition induces nonlinear changes in the stability of the contact network and in the pathways of energy dissipation, which manifest at the macroscopic scale as the statistical significance of quadratic terms.

3.4.3. Determination and Verification of Optimal Parameter Combination

In order to obtain the best repose angle parameter combination, the actual average repose angle of 40.75° is taken as the optimization objective to solve the coded equation of Group 4 in Table 8, and the constraint conditions are as follows:

$$\begin{cases} \theta(X0, X1, X2, X5) = 40.75 \\ \text{s.t.} \begin{cases} 0.14 < X0 < 0.26 \\ 0.11 < X1 < 0.23 \\ 0.02 < X2 < 0.14 \\ 0.06 < X5 < 0.22 \end{cases} \end{cases} \quad (21)$$

The optimal solution combination of the four significant factors is obtained. The surface energy of snow particles is 0.23 J/m^2, the snow–snow restitution coefficient is 0.141, the snow–snow static friction coefficient is 0.05, and the snow–membrane static friction coefficient is 0.18. The remaining non-significant factors are selected with intermediate horizontal values, that is, the snow–snow rolling friction coefficient is 0.03, the snow—membrane restitution coefficient is 0.2, and the snow–membrane rolling friction coefficient is 0.03. The solution results are simulated and verified by EDEM. Using the measurement method in Section 3.1 to calculate the repose angle, Figure 7 illustrates the analysis process and results. Polynomial fitting achieves an R^2 of 0.9976. This indicates that the measurement results are accurate and reliable.

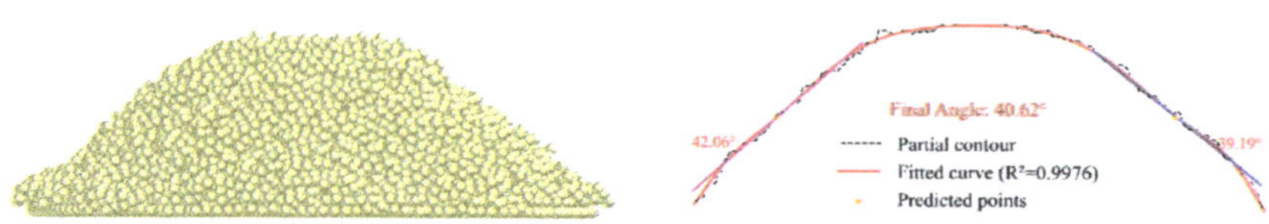

(a) Accumulation body model **(b)** Repose angle measurement results

Figure 7. Simulation result with optimal parameters.

The result shows that left and right repose angles were measured as 42.06° and 39.19°, respectively. The mean value was 40.62°, showing a 0.32% relative error compared to the actual 40.75°, confirming the reliability of the derived contact parameters.

3.4.4. Discussion on the Physical Rationality of the Optimal Parameters

The calibrated parameters are analyzed below to demonstrate their reality and physical plausibility.

Snow–snow restitution coefficient (0.141): This relatively low value indicates significant energy dissipation during collisions, which is consistent with the inelastic deformation and micro-fracture behavior of ice particles under low-velocity impact. Experimental work by Higa et al. [38] showed that for centimeter-sized ice spheres, the coefficient of

https://doi.org/10.3390/app16020610

restitution can rapidly drop below 0.3 in the "inelastic regime" where fracture occurs. Although macroscopic fracture does not occur in our simulations, the energy dissipation mechanisms involved in the numerous collisions during pile formation—such as plastic deformation—are analogous. Therefore, the value of 0.141 is physically reasonable.

Surface energy of snow particles (0.23 J/m^2): This parameter governs the adhesive force in the JKR model. For comparison, Boinovich et al. [39] reported a surface energy of approximately 0.122 J/m^2 for dry snow particles. Our calibrated value of 0.23 J/m^2 is of the same order of magnitude, confirming the reliability of our result. The discrepancy may stem from our use of a spherical particle model, which effectively accounts for some of the mechanical interlocking effects of real, irregularly shaped snow particles as increased surface adhesion.

Snow–snow static friction coefficient (0.05): Classical work by Israelachvili [40] documents a typical static friction coefficient of about 0.10 for ice–ice contacts, and the sliding friction coefficient is 0.03. The 0.05 calibrated in this study is in between the two, with a minor difference, and is physically realistic.

Snow–membrane static friction coefficient (0.18): The coefficient of friction between typical polymer materials and snow usually falls within the range of 0.1 to 0.3 [41–43]. Our experimentally calibrated value of 0.18 for the snow–membrane interface is consistent with these known frictional characteristics of snow–polymer contacts.

Regarding the three parameters identified as non-significant in the Plackett–Burman test: The snow–snow rolling friction coefficient of 0.03 aligns with the low rolling resistance typically exhibited by dry granular materials like snow, where energy dissipation via rolling is considerably smaller than that via static or sliding friction. The snow–membrane restitution coefficient of 0.2 indicates a moderately inelastic collision, which is physically reasonable for impacts between snow particles and a deformable polymer membrane, as energy is dissipated through material damping and plastic deformation at the contact point. The snow–membrane rolling friction coefficient of 0.03 is a low value, consistent with the minimal rolling resistance expected on a smooth membrane surface.

Table 11 summarizes the calibrated contact parameters obtained in this study and compares them with the physically reasonable ranges reported in the literature on snow particle properties, together with the corresponding data sources and references. The comparison demonstrates that all seven calibrated parameters fall within well-established physical ranges and exhibit clear physical significance, confirming their suitability as a reliable parameter set for subsequent discrete element simulations.

Table 11. Comparison between the calibrated DEM contact parameters obtained in this study and the ranges reported in the literature.

Parameter	Calibrated Value	Reported Value	References
Snow–snow restitution coefficient	0.141	0.1–0.3	[38]
Surface energy of snow particles	0.23 J/m^2	0.122 J/m^2	[39]
Snow–snow static friction coefficient	0.05	0.03–0.1	[40]
Snow–membrane static friction coefficient	0.18	0.1–0.3	[41–43]
Snow–snow rolling friction coefficient	0.03	0.03–0.1	[40]
Snow–membrane restitution coefficient	0.2	0.1–0.3	[41–43]
Snow–membrane rolling friction coefficient	0.03	0.01–0.05	[41–43]

It should be noted that the inverse identification of DEM contact parameters based on a single macroscopic response may inherently involve a degree of non-uniqueness. The objective of this study is not to seek a mathematically unique solution, but to obtain a

https://doi.org/10.3390/app16020610

physically consistent and statistically reliable parameter set that can accurately reproduce the macroscopic flow behavior of snow particles in discrete element simulations.

4. Conclusions

This study establishes a physically grounded and statistically robust calibration framework for discrete element modeling of snow, and the main findings can be summarized as follows:

(1) Through a Plackett–Burman (P–B) screening design, four contact parameters were identified as statistically significant factors governing the angle of repose of snow particles, namely the surface energy of snow particles, the snow–snow static friction coefficient, the snow–snow restitution coefficient, and the snow–membrane static friction coefficient. This result clarifies the dominant microscopic parameters controlling macroscopic pile stability under the investigated conditions.

(2) An image-based angle-of-repose measurement method was proposed based on polynomial fitting of the complete pile outline. Unlike conventional approaches relying on limited local slopes, the proposed method utilizes the full contour information of the granular pile, thereby improving the robustness and accuracy of repose angle estimation in the presence of geometric irregularities and particle-scale fluctuations.

(3) A response surface optimization framework incorporating Q–Q plot reference line construction and residual-based outlier treatment was established. By iteratively correcting error-prone data points through repeated simulations rather than direct exclusion, the reliability and statistical consistency of the response surface model were significantly improved.

(4) Based on the optimized second-order regression model, a physically consistent set of DEM contact parameters was obtained. The calibrated values included a snow surface energy of 0.23 J/m^2, a snow–snow restitution coefficient of 0.141, a snow–snow static friction coefficient of 0.05, and a snow–membrane static friction coefficient of 0.18, while intermediate values were retained for the remaining parameters. Using this parameter set, the simulated angle of repose reached 40.62°, corresponding to a relative error of only 0.32% compared with the experimental reference value of 40.75°. These results demonstrate that the proposed calibration strategy is both statistically sound and physically meaningful for reproducing the macroscopic flow behavior of snow.

In future studies, the calibrated contact parameters will be applied to discrete element method (DEM) simulations of snow removal robots operating on membrane structures. The DEM framework will be used to quantitatively predict the snow removal performance of the robot, as well as the accumulation, adhesion, and flow behavior of snow during the removal process. Based on these simulation results, key components of the snow removal robot—such as the geometry of the snow pusher, the design of the snow-blowing pipeline, and the overall system layout—can be systematically optimized.

Author Contributions: Conceptualization, F.H.; Methodology, J.D., F.Z. and F.H.; Software, J.D.; Validation, J.D., F.Z. and X.M.; Investigation, J.D.; Data curation, J.D., F.Z. and F.H.; Writing—original draft, J.D. and X.M.; Writing—review & editing, J.D., F.Z. and X.M.; Visualization, X.M.; Supervision, F.Z., F.H. and X.M.; Project administration, F.Z. and F.H.; Funding acquisition, F.Z., F.H. and X.M. All authors have read and agreed to the published version of the manuscript.

Funding: This work was supported by a project grant from Hebei Province central guide local science and technology development fund project (246Z1808G); Major science and technology support project of Hebei Province (242Q1802Z); Science and Technology Innovation Team Project of Shijiazhuang City (248790186A).

Institutional Review Board Statement: Not applicable.

https://doi.org/10.3390/app16020610

Informed Consent Statement: Not applicable.

Data Availability Statement: The original contributions presented in this study are included in the article. Further inquiries can be directed to the corresponding author.

Conflicts of Interest: The authors declare no conflicts of interest.

References

1. Zhang, A.Y.; He, J.H. The impact of climate change on crop growth and countermeasures. *Rural Sci. Technol. Exp.* **2025**, *5*, 193–195. [CrossRef]
2. Xiang, Y.; Li, Y.; Yu, P.B.; Wang, Y.; Tang, Y.; Xue, F.; Zhuang, X.N.; Liu, J. Innovative development and engineering design application optimization of grain air-film warehouses. *Grain Storage* **2024**, *53*, 1–5+16.
3. Li, M.S.; Wang, Y.J.; Xie, S.Y.; Liu, Y.F.; Chen, X.Z.; Li, X.; Liu, W. Simulation and experimental analysis of the operational performance of deep-loosening machines based on discrete element method. *Trans. Chin. Soc. Agric. Eng.* **2024**, *40*, 81–90. [CrossRef]
4. Johnson, J.B.; Hopkins, M.A. Identifying microstructural deformation mechanisms in snow using discrete-element modeling. *J. Glaciol.* **2005**, *51*, 432–442. [CrossRef]
5. Bobillier, G.; Bergeld, B.; Capelli, A.; Dual, J.; Gaume, J.; Van Herwijnen, A.; Schweizer, J. Micromechanical modeling of snow failure. *Cryosphere* **2020**, *14*, 39–49. [CrossRef]
6. Hagenmuller, P.; Chambon, G.; Naaim, M. Microstructure-based modeling of snow mechanics: A discrete element approach. *Cryosphere* **2015**, *9*, 1969–1982. [CrossRef]
7. Zhao, Z.H.; Wu, M.L.; Xie, S.P.; Luo, H.F.; Li, P.C.; Zeng, Y.; Jiang, X.H. Discrete element simulation parameter calibration of soil-rice rotation and its rotary tillage trajectory analysis. *Trans. Chin. Soc. Agric. Eng.* **2024**, *40*, 72–82. [CrossRef]
8. Gao, X.Y.; Wu, P.; Li, X.Z.; Du, H.J.; Wang, W.C.; Wan, Q.H. Research on the repose angle test of granular feed based on MATLAB. *J. Agric. Mech. Res.* **2024**, *46*, 173–179. [CrossRef]
9. Zhang, H.Z.; Liu, Z.M.; Jiang, W.L.; Zhang, X.; Wu, Y. Discrete element parameter calibration of irregular metal powder flow. *Metall. Mater.* **2024**, *44*, 22–24.
10. Chen, S.; Jiang, L.J.; Lin, X.S.; Tang, X.M.; Liu, X.M.; Zhang, L. Calibration of contact parameters for flour particles based on the discrete element method. *Trans. Chin. Soc. Agric. Eng.* **2024**, *40*, 69–76. [CrossRef]
11. Zou, Y.; Tang, T.; Gao, Z.C.; Qiao, Z.D.; Hu, Y.B. Discrete element parameter calibration of quicklime powder based on particle scaling theory. *China Powder Technol.* **2023**, *29*, 81–91. [CrossRef]
12. Fierz, C.; Armstrong, R.L.; Durand, Y.; Etchevers, P.; Greene, E.; McClung, D.M.; Nishimura, K.; Satyawali, P.K.; Sokratov, S.A. *The International Classification for Seasonal Snow on the Ground*; IACS Contribution No. 1; UNESCO-IHP: Paris, France, 2009; pp. 4–13.
13. Wang, E.L.; Fu, X.; Han, H.W.; Liu, X.C.; Xiao, Y.; Leng, Y.P. Study on the mechanical properties of compacted snow under uniaxial compression and analysis of influencing factors. *Cold Reg. Sci. Technol.* **2021**, *182*, 103215. [CrossRef]
14. Zermatten, E.; Schneebeli, M.; Arakawa, H.; Steinfeld, A. Tomography-based determination of porosity, specific area and permeability of snow and comparison with measurements. *Cold Reg. Sci. Technol.* **2014**, *97*, 33–40. [CrossRef]
15. Lieblappen, R.; Fegyveresi, J.M.; Courville, Z.; Albert, D.G. Using ultrasonic waves to determine the microstructure of snow. *Front. Earth Sci.* **2020**, *8*, 34. [CrossRef]
16. Arakawa, H.; Izumi, K.; Kawashima, K.; Kawamura, T. Study on quantitative classification of seasonal snow using specific surface area and intrinsic permeability. *Cold Reg. Sci. Technol.* **2009**, *59*, 163–168. [CrossRef]
17. Duan, J.; Guo, H.; Hu, J.R.; Zhou, X.; Wu, X.; Chen, B.J. Research advances on observation and shape classification of natural snow and ice crystal particles. *Acta Meteorol. Sin.* **2023**, *81*, 685–701. [CrossRef]
18. Eidevåg, T.; Abrahamsson, P.; Eng, M.; Rasmuson, A. Modeling of dry snow adhesion during normal impact with surfaces. *Powder Technol.* **2020**, *361*, 1081–1092. [CrossRef]
19. Eidevåg, T.; Thomson, E.S.; Kallin, D. Angle of repose of snow: An experimental study on cohesive properties. *Cold Reg. Sci. Technol.* **2022**, *194*, 103470. [CrossRef]
20. Zhang, X.L.; Wang, H.D.; Xiao, P.F.; Zheng, Z.X.; Feng, X.Z. Electronic atlas of spatiotemporal distribution of snow characteristics in China. *China Sci. Data* **2022**, *7*, 24–37. [CrossRef]
21. Choi, Y.B.; Kim, R.W.; Lee, I.B. Numerical analysis of snow distribution on greenhouse roofs using CFD–DEM coupling method. *Biosyst. Eng.* **2024**, *237*, 196–213. [CrossRef]
22. Johnson, K.L.; Kendall, K.; Roberts, A.D. Surface energy and the contact of elastic solids. *Proc. R. Soc. A-Math. Phys. Eng. Sci.* **1971**, *324*, 301–313. Available online: https://www.jstor.org/stable/78058 (accessed on 3 January 2026). [CrossRef]
23. Robert, R. Why Is Ice Slippery? *Phys. Today* **2005**, *12*, 50–55. [CrossRef]

 https://doi.org/10.3390/app16020610

24. Zhao, T.T.; Feng, Y.T. Accurate scaling and coarse-graining discrete element methods for large-scale granular systems. *Chin. J. Comput. Mech.* **2022**, *39*, 365–372. [CrossRef]
25. Wang, Q.Z.; Yang, M.; Xiang, J.T.; Zhang, Q.L.; Zhang, W.J.; Li, X.J.; Hu, J.M. Calibration of discrete element parameters for seabuckthorn tea hairs based on particle scaling theory. *Agric. Res. Arid Areas* **2024**, *42*, 284–292. [CrossRef]
26. Coetzee, C.J. Review: Calibration of the discrete element method. *Powder Technol.* **2017**, *310*, 104–142. [CrossRef]
27. Zhang, H.X. Response Analysis of Pneumatic Membrane Structures Under Wind and Rain Loads. Master's Thesis, Hebei University of Science and Technology, Shijiazhuang, China, 2019.
28. Liu, Q. Research on Snow Removal of Membrane Structures Based on Thermal Melting Method. Master's Thesis, Harbin Institute of Technology, Harbin, China, 2022.
29. Washino, K.; Chan, E.L.; Miyazaki, K.; Tsuji, T.; Tanaka, T. Time step criteria in DEM simulation of wet particles in viscosity dominant systems. *Powder Technol.* **2016**, *302*, 100–107. [CrossRef]
30. Lommen, S.; Schott, D.; Lodewijks, G. DEM speedup: Stiffness effects on behavior of bulk material. *Particuology* **2014**, *12*, 107–112. [CrossRef]
31. Li, M.; He, X.; Zhu, G.; Liu, J.; Gou, K.; Wang, X. Modeling and Parameter Calibration of Morchella Seed Based on Discrete Element Method. *Appl. Sci.* **2024**, *14*, 11134. [CrossRef]
32. Vázquez-Martín, S.; Kuhn, T.; Eliasson, S. Shape dependence of snow crystal fall speed. *Atmos. Chem. Phys.* **2021**, *21*, 7545–7565. [CrossRef]
33. Myers, R.H.; Montgomery, D.C.; Anderson-Cook, C.M. *Response Surface Methodology: Process and Product Optimization Using Designed Experiments*, 4th ed.; John Wiley & Sons: Hoboken, NJ, USA, 2016; pp. 273–307.
34. Fang, J.T. Comparative Study on Experimental Design and Model Estimation in Response Surface Methodology. Master's Thesis, Tianjin University, Tianjin, China, 2011.
35. Weine, E.; McPeek, M.S.; Abney, M. Application of equal local levels to improve Q-Q plot testing bands with R package qqconf. *J. Stat. Softw.* **2023**, *106*, 1–27. [CrossRef]
36. Li, Y.M. Methods for normality test of data. *J. Huaihua Univ.* **2015**, *34*, 81–82. [CrossRef]
37. Rakhi; Gupta, B.; Lamba, S.S. An efficient local outlier detection approach using kernel density estimation. *Franklin Open* **2024**, *8*, 100162. [CrossRef]
38. Higa, M.; Arakawa, M.; Maeno, N. Size dependence of restitution coefficients of ice in relation to collision strength. *Icarus* **1998**, *133*, 310–320. [CrossRef]
39. Boinovich, L.; Emelyanenko, A. Experimental determination of the surface energy of polycrystalline ice. *Dokl. Phys. Chem.* **2015**, *459*, 198–202. [CrossRef]
40. Israelachvili, J.N. *Intermolecular and Surface Forces*, 3rd ed.; Academic Press: Waltham, MA, USA, 2011; pp. 471–474.
41. Klapproth, C.; Kessel, T.M.; Wiese, K.; Wies, B. An advanced viscous model for rubber-ice-friction. *Tribol. Int.* **2016**, *99*, 169–181. [CrossRef]
42. Bäurle, L.; Kaempfer, T.U.; Szabó, D.; Spencer, N.D. Sliding friction of polyethylene on snow and ice: Contact area and modeling. *Cold Reg. Sci. Technol.* **2007**, *47*, 276–289. [CrossRef]
43. Tan, T.; Xing, C.; Tan, Y.Q. Rubber friction on icy pavement: Experiments and modeling. *Cold Reg. Sci. Technol.* **2020**, *174*, 103022. [CrossRef]

https://doi.org/10.3390/app16020610

applied sciences

MDPI

Article

Bioinspired Kirigami Structure for Efficient Anchoring of Soft Robots via Optimization Analysis

Muhammad Niaz Khan [1,2], Ye Huo [1,2], Zhufeng Shao [1,2,*], Ming Yao [1,2] and Umair Javaid [3]

1 State Key Laboratory of Tribology, Tsinghua University, Beijing 100084, China; mniazkhan22@mails.tsinghua.edu.cn (M.N.K.); huoy21@mails.tsinghua.edu.cn (Y.H.)
2 Beijing Key Laboratory of Transformative High-End Manufacturing Equipment and Technology, Tsinghua University, Beijing 100084, China
3 School of Robotics, Ningbo University of Technology, Ningbo 315211, China
* Correspondence: shaozf@mail.tsinghua.edu.cn

Abstract: Kirigami-inspired geometries offer a lightweight, bioinspired strategy for friction enhancement and anchoring in soft robotics. This study presents a bioinspired kirigami structure designed to enhance the anchoring performance of soft robotic systems through systematic geometric and actuation parameter optimization. Drawing inspiration from the anisotropic friction mechanisms observed in reptilian scales, we integrated linear, triangular, trapezoidal, and hybrid kirigami cuts onto flexible plastic sheets. A compact 12 V linear actuator enabled cyclic actuation via a custom firmware loop, generating controlled buckling and directional friction for effective locomotion. Through experimental trials, we quantified anchoring efficiency using crawling distance and stride metrics across multiple cut densities and actuation conditions. Among the tested configurations, the triangular kirigami with a 4 × 20 unit density on 100 μm PET exhibited the most effective performance, achieving a stride efficiency of approximately 63% and an average crawling speed of ~47 cm/min under optimized autonomous operation. A theoretical framework combining buckling mechanics and directional friction validated the observed trends. This study establishes a compact, tunable anchoring mechanism for soft robotics, offering strong potential for autonomous exploration in constrained environments.

Keywords: kirigami structure; bionic design; anisotropic friction; optimization analysis; soft robotics

1. Introduction

The field of soft robotics has emerged as a transformative domain in automation, offering compliance, adaptability, and resilience to navigate unstructured and constrained environments—capabilities that rigid robotic systems often lack [1–3]. Among nature's most efficient movers, snakes exhibit remarkable terrain adaptability, navigating tight crevices or uneven surfaces using a repertoire of locomotion strategies such as rectilinear progression and lateral undulation [4–6]. Central to their locomotion efficiency is the anisotropic friction generated by ventral scales, which provide directional anchoring and propulsion with minimal energy expenditure [4,7]. These scales interlock with ground asperities and generate higher resistance to backward or lateral motion than forward, as experimentally shown by Hu et al. [4], who quantitatively modeled how belly scale anisotropy enables effective slithering on flat terrain.

Drawing inspiration from biomechanical principles, researchers have sought to emulate this anchoring capability in soft robotic platforms, targeting applications such as confined space navigation, biomedical systems, and terrain-adaptive exploration where traditional robots are not an appropriate option [8–11]. One promising approach involves the use of kirigami, the ancient art of patterned cuts in thin sheets—reimagined as a tool to engineer deformable surfaces with tunable mechanical and frictional properties to engage the surface in a scale-like fashion [12–14].

The morphological diversity found in reptilian scales offers a bioinspired template for the design of kirigami skins capable of controlled deformation and surface interaction. In this study, five distinct kirigami patterns are explored to optimize anchoring behavior: linear cuts that provide minimal frictional interaction but maintain structural stability; triangular cuts inspired by the sharp-edged ventral scales of Puff Adders, enhancing anisotropic friction and grip; trapezoidal patterns derived from the whiptail lizard, offering a balance between flexibility and mechanical support; and two hybrid variants, linear–triangular and linear–trapezoidal, that combine stability with directional anchoring features [6,15–18]. These structures enable tunable anisotropic friction under compression, echoing the engagement–release anchoring mechanics observed in snakes during slithering and concertina locomotion [19]. Prior implementations of kirigami-based soft robotic systems have commonly relied on pneumatic actuation to induce the necessary buckling deformation for locomotion. However, these systems often suffer from bulk, noise, and the need for sealed chambers and external pumps, limiting their compactness and autonomy [20,21]. This study utilizes a permanent magnet (PM) linear actuator as the actuation source. This actuator provides a silent, energy-efficient, and compact solution, enabling controlled axial compression and repeatable deformation cycles of the kirigami skin [22,23].

Overall, this work is to optimize the anchoring performance of a kirigami-based soft robotic crawler through systematic variation of both structural and actuation parameters. Anchoring efficiency is assessed via crawling distance, stride efficiency, and net propulsive force under varying conditions of cut geometry, PET sheet thickness, cut density, actuator stroke, and actuation speed. The combination of experimental testing with analytical modeling allows for a comprehensive investigation of how geometric and mechanical factors contribute to effective directional anchoring. To support the design process, a theoretical model is developed that integrates classical buckling mechanics with a friction-based analytical framework, enabling prediction of pop-up height and anisotropic friction forces. This dual perspective offers not only validation of observed results but also a foundation for further optimization of kirigami-based anchoring systems.

Further, this paper is structured as follows: Section 2 presents the design principles, fabrication methodology, and experimental setup for the kirigami-skinned robot. Section 3 provides detailed results from parametric studies and theoretical modeling. Section 4 discusses the implementation of an autonomous crawling system and its performance metrics. Finally, the conclusion summarizes key findings and outlines future research directions in sensor integration and terrain-adaptive anchoring strategies.

2. Design, Composer, and Experimental Setup

This section outlines the design principles, fabrication methods, and control strategies behind the kirigami-based soft robotic crawler, integrating bio-inspired kirigami structures with a PM linear actuator to optimize locomotion efficiency. We detail the kirigami skin's geometry, material selection, actuator integration, experimental protocols, and a theoretical framework predicting performance, all validated by experimental outcomes.

2.1. Kirigami Structure Design

The kirigami structures were fabricated with arrays of patterns into polyester plastic sheets (PET) ($E = 4.33$ GPa, $\nu = 0.4$), chosen for their balance of flexibility and durability. The intricate kirigami patterns enabled the creation of designs tailored for specific frictional and locomotive roles: linear cuts as a baseline with high stability but minimal friction; triangular cuts, inspired by the Puff Adder's sharp-edged scales (~5–8 mm [7,18,24]), for enhanced anisotropic friction and anchoring; and trapezoidal cuts, inspired by whiptail lizard ventral scales (~4–7 mm [25,26]), for balanced flexibility and support (Figure 1a). Prior biological studies have shown that snakes exploit directional friction coefficients (forward vs. backward vs. lateral) to optimize motion, often producing passive anchoring at specific body segments through contact angle modulation and scale placement [4]. This behavior has been quantitatively linked to scale geometry and surface interaction. Our asymmetrical kirigami cuts are explicitly designed to emulate these biomechanical strategies by deforming asymmetrically under compression, thereby maximizing anchoring in the desired direction of motion. Proposed design parameters for kirigami pattern analysis for optimization included a lattice unit defined by width $L = 2l\cos(\pi/6)$ and height $H = 2l\sin(\pi/6)$, where l is the cut length and δ is the hinge width separating cuts at an angle $\gamma = \pi/3$ (Figure 1b). Fabrication involved cutting with the Silhouette Cameo 5 (Silhouette America, Inc., Lehi, UT, USA), producing expandable kirigami structures (Figure 1e,f).

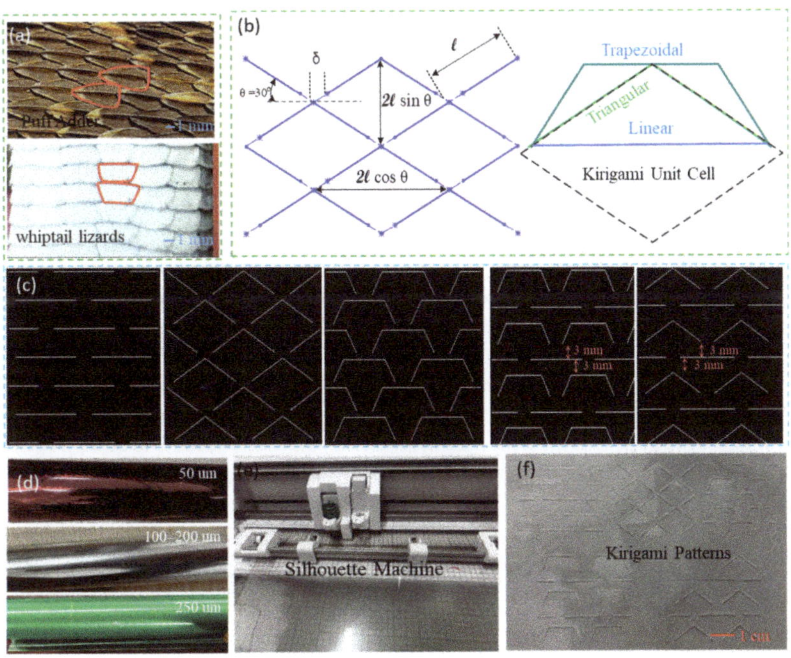

Figure 1. Bioinspired kirigami design and fabrication workflow. (**a**) Triangular and trapezoidal scale patterns of the Puff Adder and whiptail lizards used as biological inspiration [15,24]. (**b**) Geometric definition of kirigami unit cells, showing linear, triangular, and trapezoidal types. (**c**) CAD designs of kirigami patterns, including linear, triangular, trapezoidal, and hybrid cuts. (**d**) PET sheets with different thicknesses selected to tune flexibility and stiffness. (**e**) Fabrication of kirigami sheets using a Silhouette Cameo 5 digital cutting machine. (**f**) Resulting physical kirigami samples displaying clean, high-precision cuts across different patterns.

2.2. Composer of the Kirigami-Skinned Crawlers for Parameters Analysis

To systematically evaluate the functional impact of kirigami structures on locomotion, a modular soft robotic crawler was composed, integrating mechanical, structural, and electronic subsystems. The crawler comprises a permanent magnet (PM) linear actuator, a kirigami-skinned silicone body, and a microcontroller-based control unit, designed to enable precise and repeatable testing under varied experimental conditions. The PM linear actuator (Model: LA-T8, GoMotorWorld, Shenzhen, China; 12 V, 60 N thrust, 15 mm/s rated speed, 30 mm strokewas selected for its compactness, silent operation, and higher precision compared to traditional pneumatic actuators [23,27]. The actuator provides a maximum stroke length of 30 mm, sufficient for evaluating different deformation regimes and crawling behaviors (Figure 2a). It is mechanically coupled to a compliant soft body formed from Ecoflex 00-30, silicone elastomer ((Smooth-On, Inc., Macungie, PA, USA; shear modulus $\mu_a \approx 30$ kPa), cast into a 1-millimeter-thick sheet, and degassed to eliminate voids. This flexible substrate offers low stiffness and high stretchability, allowing seamless integration with kirigami layers. The kirigami skin is fabricated from PET sheets (Changzhou Huisu Qinye Plastic Group, China) with kirigami cut patterns and is firmly bonded to the silicone layer from both sides, which ensures consistent attachment and mechanical transfer (Figure 2b). Upon actuation, the sheet undergoes buckling-induced pop-up deformation, translating the actuator's linear motion into anisotropic surface contact and friction-driven propulsion. The closed configuration of the actuator maintains tension on the skin, which rebounds during retraction, facilitating the cyclic crawling motion.

A microcontroller-based control circuit governs the actuation cycle of the composed crawler. The control system features an Arduino and an LN298 motor driver, powered by dual 12 V lithium batteries (Kuongshun Electronic Limited, Guangzhou, China), delivering both driving power and regulated logic voltage. The actuator's bidirectional movement is controlled via a polarity-switching algorithm, enabling fine-tuned extension and retraction sequences. A standard test consisted of five actuation cycles at 15 mm/s extension speed, unless otherwise varied, with displacement tracked across a polyurethane (PU) foam substrate that simulates compliant terrain (Figure 2c,d). To further understand frictional dynamics during locomotion, forward and backward friction forces were independently measured using a J14014 mechanical sensor (Model: Phyphox, Xiangsheng Science and Education Equipment Co., Ltd., Guang'an, China). In a separate experimental setup, the crawler was placed on PU foam while the sensor recorded resistance during controlled to-and-from motion (Figure 2e,f). This allowed for the quantification of anisotropic friction generated by various kirigami configurations, critical for optimizing propulsion mechanics. To systematically optimize the locomotion efficiency of the kirigami-based soft robotic crawler, a series of controlled experiments were conducted. Each test was designed to evaluate the influence of specific structural or actuation parameters on crawling performance. Each experiment was repeated five times to ensure repeatability, and the mean crawling distance along with standard deviation was recorded to quantify the effect of pattern geometry on traction, deformation behavior, and net displacement. The overarching objective was to understand the complex interplay between kirigami mechanics, surface interaction, and actuation dynamics. The following subsections describe each experiment and its purpose:

1. Kirigami Structure Type vs. Crawling Distance: This experiment investigates how different kirigami geometries affect anisotropic friction and locomotion. Inspired by the morphological diversity of snake scales, five distinct patterns were tested: linear, trapezoidal, triangular, hybrid linear–triangular, and hybrid linear–trapezoidal. The primary aim was to analyze how each pattern's deformation mechanics and surface contact profile contribute to directional friction and, thus, crawling performance. By

comparing the average crawling distance over five actuation cycles, this study aims to identify the most efficient pattern for generating propulsive force (Figure 1c).

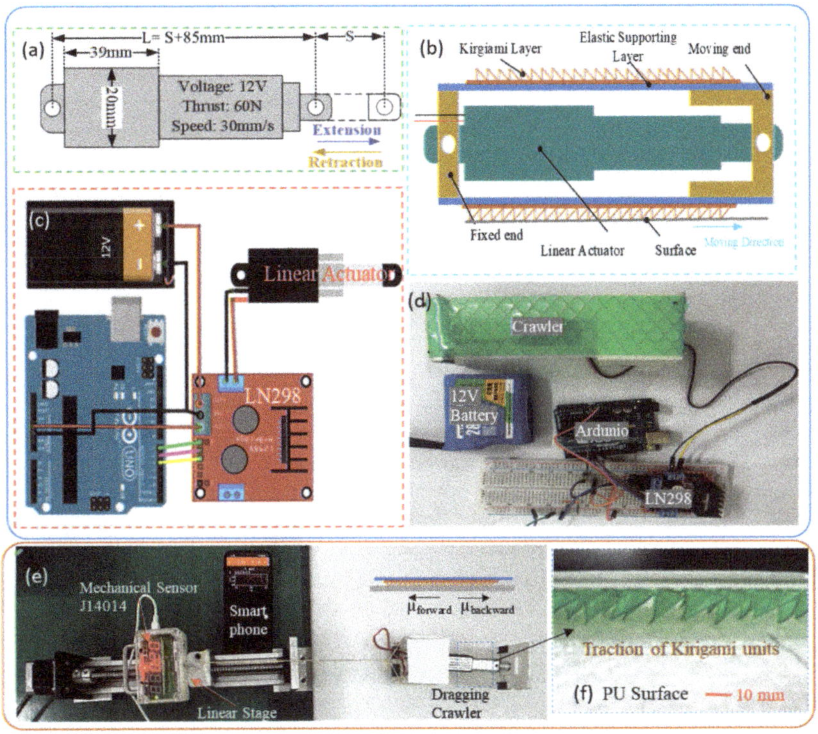

Figure 2. Design schematic, electronic integration, and experimental setup of the kirigami-based soft crawling robot. (**a**) Schematic of the 12 V PM linear actuator (60 N thrust, 30 mm/s speed) with labeled extension/retraction strokes and dimensions. (**b**) Mechanical layout of the crawler showing bilaterally symmetric kirigami skins oriented for directional locomotion. (**c**) Wiring diagram of the control system comprising an Arduino Uno, LN298 motor driver, and 12 V battery (See text for manufacturer details), used to actuate the crawler. (**d**) Fully assembled prototype showing the integrated crawler, power unit, and control board. (**e**) Friction measurement platform using a linear stage and digital mechanical sensor, with a smartphone for synchronized data logging. (**f**) Close-up view of triangular kirigami units interacting with a polyurethane (PU) substrate.

2. PET Sheet Thickness vs. Crawling Distance: In this test, the effect of substrate flexibility and stiffness on locomotion was examined by using PET sheets of varying thicknesses: 50 μm (red), 100 μm, 150 μm, 200 μm (transparent), and 250 μm (green). The experiment evaluates the average crawling distance to determine the optimal balance between mechanical compliance and anchoring effectiveness, both of which are critical for reliable propulsion (Figure 1d).
3. Kirigami Cuts Density vs. Crawling Distance: To assess the impact of surface coverage and unit repetition on locomotion performance, we analyzed five different kirigami cut densities, i.e., 2 × 14, 3 × 15, 4 × 20, 5 × 28, and 6 × 28, implemented over a standardized contact area of 51 mm × 140 mm. In this notation, each density value M×N represents a grid of M cut units along the longitudinal (lengthwise) axis and N units along the transverse (widthwise) axis, covering the entire kirigami sheet.

This structured arrangement modulates the number, spacing, and distribution of pop-up units, thereby influencing inter-unit mechanical interference, surface anchoring behavior, and crawling performance.

4. By systematically varying the cut densities, we evaluated how increasing the number of units affects deformation dynamics and effective engagement with the ground surface. This approach enabled identification of an optimal density configuration that maximizes propulsion while minimizing performance losses due to structural interference.
5. Stroke Length vs. Crawling Distance: This test explores how actuator stroke amplitude influences the mechanical engagement of the kirigami skin with the substrate. The following four actuation protocols were examined: 10 mm (St10), 20 mm (St20), 30 mm (St30), and 30 mm with a 1-second hold before retraction (St30h1s). By measuring the crawling distance over five actuation cycles, the experiment evaluates how increased extension—combined with dwell time—affects unit deformation, surface interaction, and net displacement.
6. Actuation Speed vs. Crawling Distance: This experiment evaluates how different extension/retraction speeds—10/10, 15/15, 20/20, and 30/30 mm/s—influence crawling performance. Actuation speed directly affects the contact time and engagement quality between kirigami structures and the surface. By comparing average crawling distances, the test aims to determine a speed regime that optimally balances actuation dynamics with surface interaction mechanics for efficient locomotion.

2.3. Theoretical and Analytical Design

To guide the design and optimization of the kirigami-based soft robotic crawler, we developed a theoretical framework to predict the pop-up height (h) and anisotropic friction properties of the kirigami skin, integrating buckling mechanics with tribological analysis. This model was applied to a polyethylene terephthalate (PET) sheet ($E = 4.33$ GPa, $\nu = 0.4$) and validated using experimental data to quantify how cut geometry and actuation translate into propulsive forces driving locomotion. The framework provides a predictive tool for design refinement, with predictions for kirigami design configurations to optimize traction features.

2.3.1. Pop-Up Height Analysis

The pop-up height of the kirigami skin emerges from localized buckling triggered by axial compression applied via a PM linear actuator with a 30 mm stroke. Drawing upon classical buckling theory for thin elastic sheets [13,21,28], the pop-up height is evaluated by the relation:

$$h = kt\sqrt{\frac{\epsilon}{\epsilon_c}} \tag{1}$$

where t is the sheet thickness (100 μm); $\epsilon = \Delta L / L_0 = 0.214$ is the applied compressive strain; ϵ_c is the critical buckling strain; and k is a geometric amplification factor determined via experimental calibration. Assuming $l = 10$ mm with $L_{\text{cut}} = 9$ mm, $t = 0.1$ mm, and $\nu = 0.4$ yields. The critical strain for uniaxial buckling is given by:

$$\epsilon_c = \frac{\pi^2}{12(1-\nu^2)}\left(\frac{t}{L_{\text{cut}}}\right)^2 \tag{2}$$

Substituting values with an initial geometric factor $k = 1.0$ yields: $\epsilon_c \approx 1.207 \times 10^{-4}$ and $h \approx 4.21$ mm. Experimental validation across six measured locations (Figure 3b) evaluated via ImageJ (version 1.54 g) for samples with $l = 10$ mm, $L_{\text{cut}} = 9$ mm, yielded average pop-up heights of 8.99 mm for triangular cuts and 7.85 mm for trapezoidal cuts.

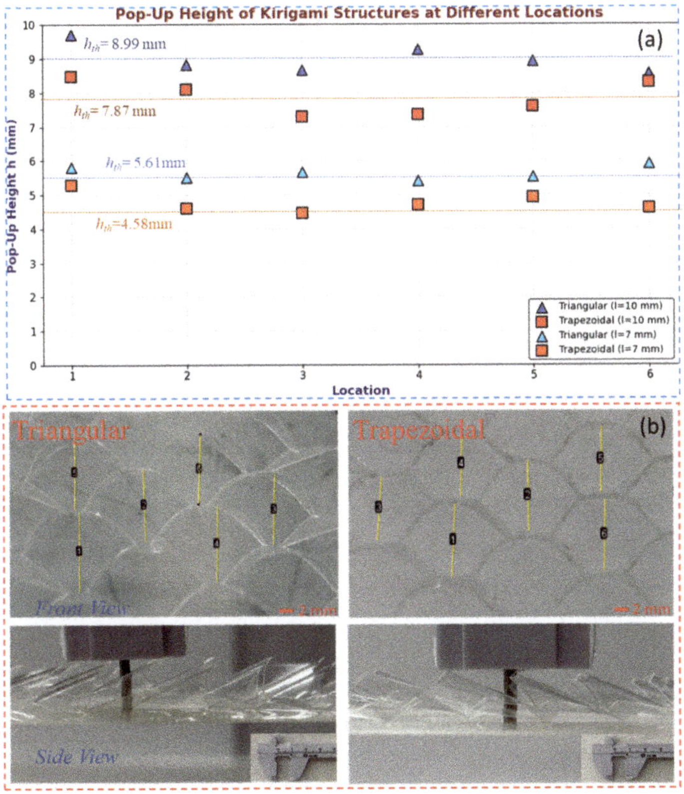

Figure 3. Quantitative and visual analysis of the pop-up height of triangular and trapezoidal kirigami structures. (**a**) Plot showing the measured pop-up heights (h_p) at six locations (labeled 1–6) across each kirigami pattern for both triangular and trapezoidal cuts with cut lengths of 10 mm and 7 mm. (**b**) Front and side views of the kirigami structures during actuation. Yellow lines represent pop-up heights extracted using ImageJ image processing and correspond to the data shown in (**a**).

To adjust the theoretical predictions with experimental data, the factor k was calibrated to 2.14 for triangular patterns and 1.87 for trapezoidal ones. This yielded refined predictions of $h \approx 9.00$ mm for triangular and $h \approx 7.87$ mm for trapezoidal geometries. These elevated k values reflect the additional geometric amplification introduced by kirigami patterns, which lead to more complex tent-like deformations beyond simple Euler buckling, with sharper triangular cuts exhibiting greater vertical displacement. For further comparison, when $L_{\text{cut}} = 6$ mm, the critical buckling strain increases to $\epsilon_c \approx 2.72 \times 10^{-4}$, and using the previously calibrated $k = 2.14$, the predicted pop-up height becomes approximately 5.61 mm for triangular and 4.58 mm for trapezoidal cuts. This prediction corresponds closely with experimental results for triangular and trapezoidal patterns as observed in Figure 3a, which further confirms the robustness of the model in capturing the influence of cut length and geometry on vertical deformation.

2.3.2. Anisotropic Friction Analysis

Directional locomotion in the crawler is enabled by anisotropic friction forces generated by the interaction of kirigami pop-ups with the substrate. As the actuated kirigami structures buckle upward, they form inclined flaps that generate differential friction in the forward and backward directions. This behavior was analytically modeled by expressing the effective forward and backward friction coefficients such as:

$$\mu_{forward} = \mu_0 + k_f h\rho, \quad \mu_{backward} = \mu_0 - k_f h\rho \tag{3}$$

where μ_0 is the intrinsic base friction coefficient of the PET material against the substrate; k_f is a friction amplification constant that encapsulates both the inclination angle and surface engagement; h is the pop-up height as described earlier; and ρ is the number of kirigami units of density. Using experimental results for triangular patterns at $l = 7\,\text{mm}$, configuration—where $h = 5.61\,\text{mm}$, $\rho = 0.0112$ for 4×20 kirigami units over 51 mm × 140 mm, $N = 0.58\,\text{N}$, $\mu_{\text{forward}} = 2.241$ and $\mu_{\text{backward}} = 0.776$. Based on the experimental result, the base coefficient and enhancement factors were computed as follows:

$$\mu_0 = 1.5085, \quad k_f \approx 11.66\,\text{mm}^{-1} \tag{4}$$

The corresponding net propulsive force can be calculated as follows:

$$F_{\text{prop}} = N(\mu_{\text{forward}} - \mu_{\text{backward}}) \approx 0.850\,\text{N} \tag{5}$$

This propulsive force aligns closely with experimental locomotion performance and clearly demonstrates the functional contribution of directional friction enabled by different kirigami structures (Section 3). Overall, this integrated theoretical framework provides a valuable predictive tool for rational design and iterative optimization of kirigami-based soft robotic systems.

3. Results and Analysis

3.1. *Effect of Kirigami Structure on Crawling Efficiency*

3.1.1. Experimental Conditions

To investigate the influence of kirigami pattern geometry on locomotion performance, five distinct cut configurations—linear, triangular, trapezoidal, hybrid linear–triangular, and hybrid linear–trapezoidal—were fabricated on 100 µm transparent PET sheets (Young's modulus E = 4.33 GPa, Poisson's ratio ν = 0.4). Each pattern was designed with a cut density of 3 × 14 units, uniformly distributed across a 51 mm × 140 mm area and aligned longitudinally with the direction of actuation. The crawling performance of each kirigami configuration was evaluated on a compliant polyurethane (PU) foam substrate, chosen to replicate soft, friction-sensitive terrain. A precision PM linear actuator with a 30 mm stroke and a constant extension–retraction speed of 15 mm/s was employed to drive locomotion. For each test, the crawler underwent five full actuation cycles, and the average crawling distance was recorded to quantify the effect of pattern geometry on traction, deformation behavior, and net displacement.

3.1.2. Performance Analysis

Crawling distances varied significantly across patterns (Figure 4). The linear pattern, designed as a baseline with high stability but minimal friction, achieved only 5 mm due to its isotropic friction. In contrast, the triangular pattern, inspired by the Puff Adder's sharp-edged scales for enhanced anisotropic friction, covered 65 mm. The trapezoidal pattern, inspired by whiptail lizard ventral scales for balanced flexibility and support, reached 51 mm. The hybrid linear–triangular pattern, combining stability with improved directional grip, recorded 59 mm, while the hybrid linear–trapezoidal pattern, merging friction and stability, achieved 60 mm. These results highlight the superior anisotropic friction of triangular and trapezoidal geometries, which enhance surface engagement through pop-up structures, as predicted by our buckling model. Better results of the triangular pattern stem from its ability to maximize forward friction while minimizing backward slip. The evaluated friction coefficients were $\mu_{\text{forward}} = 2.241$ and $\mu_{\text{backward}} = 0.776$, generating a propulsive force of $F_{\text{prop}} \approx 0.850\,\text{N}$. This aligns with the observed crawling distance of 65 mm (13 mm per cycle), yielding a crawling efficiency of ~43% (13 mm per 30 mm stroke). The pop-up height for this configuration was calculated as $h = 5.61$ mm, which facilitates strong surface anchoring. The trapezoidal pattern, with measured friction coefficients ($\mu_{\text{forward}} = 2.414$, $\mu_{\text{backward}} = 0.862$), produced a slightly higher propulsive force ($F_{\text{prop}} \approx 0.900\,\text{N}$), but its crawling distance of 51 mm (10.2 mm per cycle, efficiency ~34%) suggests that the broader pop-up geometry increases backward friction as well, reducing overall efficiency. The hybrid linear–triangular pattern exhibited the highest anisotropy ($\mu_{\text{forward}} = 5.000$, $\mu_{\text{backward}} = 1.897$, $F_{\text{prop}} \approx 1.800\,\text{N}$), yet covered only 59 mm (11.8 mm per cycle, efficiency ~ 39%). This indicates that while the hybrid design maximizes directional grip, factors such as pop-up stability or dynamic surface interactions may limit its performance. The hybrid linear–trapezoidal pattern achieved 60 mm (12 mm per cycle, efficiency ~40%), with $\mu_{\text{forward}} = 2.586$, $\mu_{\text{backward}} = 1.552$ and $F_{\text{prop}} \approx 0.600\,\text{N}$, reflecting a balance between friction and stability.

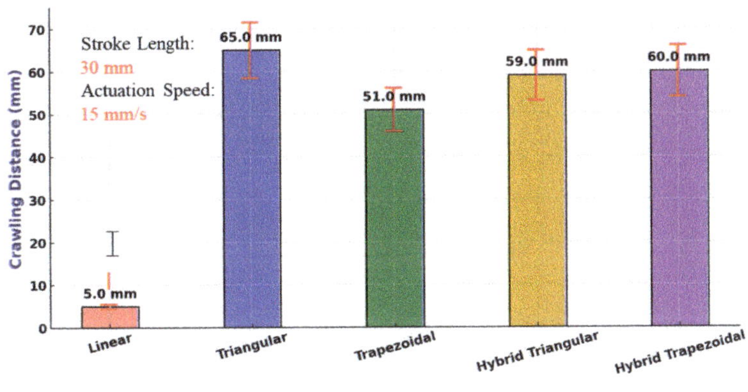

Figure 4. Crawling distance comparison of different kirigami patterns demonstrating the impact of geometry on locomotion.

The linear pattern's negligible crawling results from its lack of directional pop-up structures, leading to isotropic friction ($\mu_{\text{forward}} \approx \mu_{\text{backward}}$, $F_{\text{prop}} \approx 0\,\text{N}$). The small distance is due to minor surface interactions and actuator momentum. Compared to the linear pattern, the bioinspired designs amplify locomotion efficiency, aligning with snake scale functionality. Hybrid kirigami cuts ingeniously integrate linear cuts to introduce additional directional buckling points, enhancing the structural support of trapezoidal and triangular patterns. This synergy in hybrids explains their strong performance, with the hybrid trapezoidal pattern nearly matching the triangular pattern due to enhanced directional control. Hybrid designs illustrate the benefit of integrating linear elements with triangular or trapezoidal shapes, offering a synergistic effect that enhances both frictional engagement and structural integrity, proving particularly effective for varied terrain navigation. These results highlight the potential of kirigami cut design to enhance the efficiency of kirigami-based soft robotic locomotion. The complex geometries of the triangular and hybrid trapezoidal patterns are instrumental in achieving high-performance crawling by maximizing frictional forces and enhancing surface adaptability.

3.2. Impact of Sheet Thickness

3.2.1. Experimental Conditions

In this experiment, the influence of PET sheet thickness on the crawling performance of kirigami-patterned soft robotic crawlers is evaluated. Five different thicknesses—50 μm (red), 100 μm, 150 μm, 200 μm (transparent), and 250 μm (green)—were tested using both triangular and trapezoidal kirigami patterns. Each sample incorporated a 3 × 14 cut configuration over a fixed contact area of 51 mm × 140 mm, ensuring consistent cut density across all variants. To assess locomotion, each crawler underwent five complete extension–retraction cycles driven by a PM linear actuator with a 30 mm stroke at a constant speed of 15 mm/s on a compliant polyurethane (PU) foam substrate. The goal was to determine the optimal sheet thickness that provides an effective trade-off between material flexibility for out-of-plane deformation and mechanical stability for surface anchoring across both triangular and trapezoidal kirigami configurations.

3.2.2. Performance Analysis

Sheet thickness significantly influenced crawling distance for both triangular and trapezoidal patterns, as observed in Figure 5. The 100 μm sheet achieved the best results: 62 mm for the triangular pattern and 53 mm for the trapezoidal pattern. In comparison, the triangular pattern recorded distances of 54 mm (50 μm), 48 mm (150 μm), 44 mm (200 μm), and 41 mm (250 μm), while the trapezoidal pattern achieved 58 mm (50 μm), 42 mm (150 μm), 38 mm (200 μm), and 37 mm (250 μm). These results suggest that 100 μm offers an ideal compromise: sufficient flexibility for kirigami pop-ups and structural support for effective friction. As discussed in Section 3.1, the triangular pattern at 100 μm showed better crawling performance than trapezoidal cuts due to increased backward friction reducing overall efficiency. The 50 μm sheets, while more deformable, exhibited reduced distances, likely due to insufficient rigidity to sustain optimal contact and pop-up stability. Conversely, thicker sheets (150–250 μm) showed a consistent decline in performance for both patterns, with the 250 μm sheet achieving the lowest distances (41 mm for triangular and 37 mm for trapezoidal). This decline is attributed to increased stiffness, which limits pop-up formation and reversion post-contraction, retaining buckled shapes and diminishing anisotropic friction, as observed in Figure 5. Although the theoretical pop-up height model $h = kt\sqrt{\epsilon/\epsilon_c}$ suggests that thickness t appears in both the numerator and denominator (through ϵ_c), and thus cancels out, experimental results reveal that sheet thickness still significantly affects real-world deformation and recovery behavior. While thinner sheets offer greater compliance and flexibility, excessively thin materials (e.g., 50 μm) may lack the structural integrity needed for consistent buckling and traction. Conversely, thicker sheets (e.g., 250 μm) exhibit higher stiffness and risk of plastic deformation, reducing recovery efficiency after repeated cycles. The 100 μm PET sheet achieves a balance between flexibility and structural resilience, highlighting its suitability for kirigami-based directional frictional properties. These findings underscore material thickness as a critical design parameter, complementing kirigami geometry for locomotion efficiency, with the triangular pattern slightly showing better results than the trapezoidal pattern at optimal thickness due to its higher friction anisotropy.

3.3. Kirigami Cuts Density vs. Crawling Distance

3.3.1. Experimental Setup

To investigate the impact of kirigami structure on locomotion, a systematic experimental study was conducted to evaluate how varying the kirigami cut density influences the crawling efficiency of the soft robot. The density of the kirigami structure was modified by altering the unit lattice length (l) and adjusting the number of unit cells within the fixed area. These two parameters are inherently interrelated; a higher density is achieved by reducing the unit length while increasing the number of cuts within a fixed surface area. For this study, four distinct density configurations were fabricated using triangular and trapezoidal kirigami patterns, all constructed from 100 μm thick PET sheets. The tested configurations included 2 × 14 units (l = 10 mm), 3 × 15 units (l = 9 mm), 4 × 20 units (l = 7 mm), 5 × 28 units (l = 5 mm), and 6 × 28 units (l = 4.8 mm). Each sample was subjected to five actuation cycles on a compliant polyurethane foam substrate to simulate crawling on soft terrain.

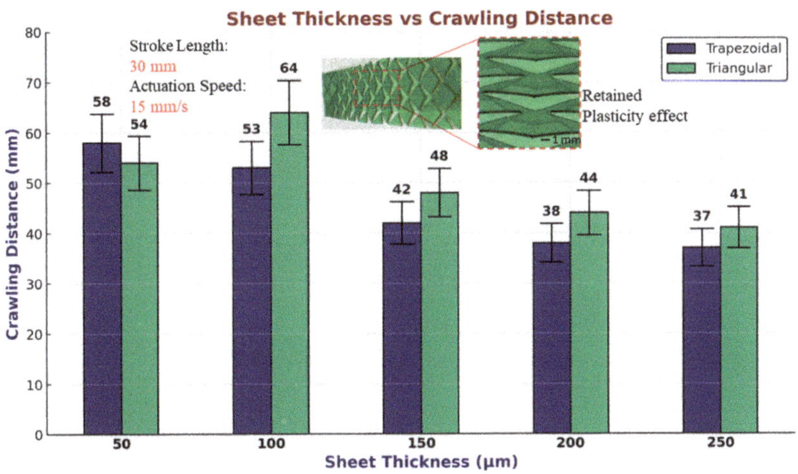

Figure 5. Average crawling distance of the robot for PET sheets with triangular and trapezoidal patterns of varying thickness.

3.3.2. Performance Analysis

Figure 6a illustrates the relationship between kirigami cut density and crawling distance for both triangular and trapezoidal patterns. The experimental results reveal a nonlinear trend, indicating that increasing cut density initially enhances crawling performance up to an optimal point, after which further increases lead to degradation. Among the tested samples, the 4 × 20 triangular kirigami configuration achieved the highest crawling distance of 74 mm, followed by 65 mm for the trapezoidal counterpart. This performance peak is attributed to an optimal balance between unit spacing and surface interaction. At this density, individual kirigami units have sufficient room to buckle vertically, achieving a favorable pop-up height while maintaining effective contact with the surface. This combination enhances anisotropic friction and enables forward propulsion. In contrast, both lower and higher densities resulted in reduced crawling performance.

For instance, the 2 × 14 configuration exhibited limited surface interaction due to fewer anchoring units and limited deformation engagement with the surface, achieving only 55 mm (triangular) and 48 mm (trapezoidal) of crawling distance. Although the 3 × 15 configuration showed modest improvement (61 mm and 54 mm), the performance dropped notably in the 5 × 28 and 6 × 28 configurations, with crawling distances decreasing to 42–44 mm and further to 34–36 mm, respectively.

The primary cause of reduced crawling performance at higher cut densities can be attributed to two interrelated factors. First, although individual kirigami units do deform across all tested configurations, the buckling height notably decreases in samples with smaller and more densely packed cuts. This lower vertical deformation limits the capacity of each unit to generate the anisotropic friction necessary for effective surface anchoring. As a result, despite undergoing buckling, these compact units fail to produce enough vertical engagement with the substrate, weakening their contribution to forward propulsion. Secondly, overcrowding from excessive unit density reduces the effective interaction area between the kirigami skin and the surface. With less spacing between adjacent units, the deformation of kirigami units may interfere with its nearby units, diminishing the structure's ability to conform and grip the surface effectively. This leads to less consistent contact and reduced anchoring force per unit during the actuation cycle. Collectively, these two effects—lower pop-up height and compromised surface engagement—cause diminished frictional force and, consequently, reduced locomotion efficiency at high cut densities.

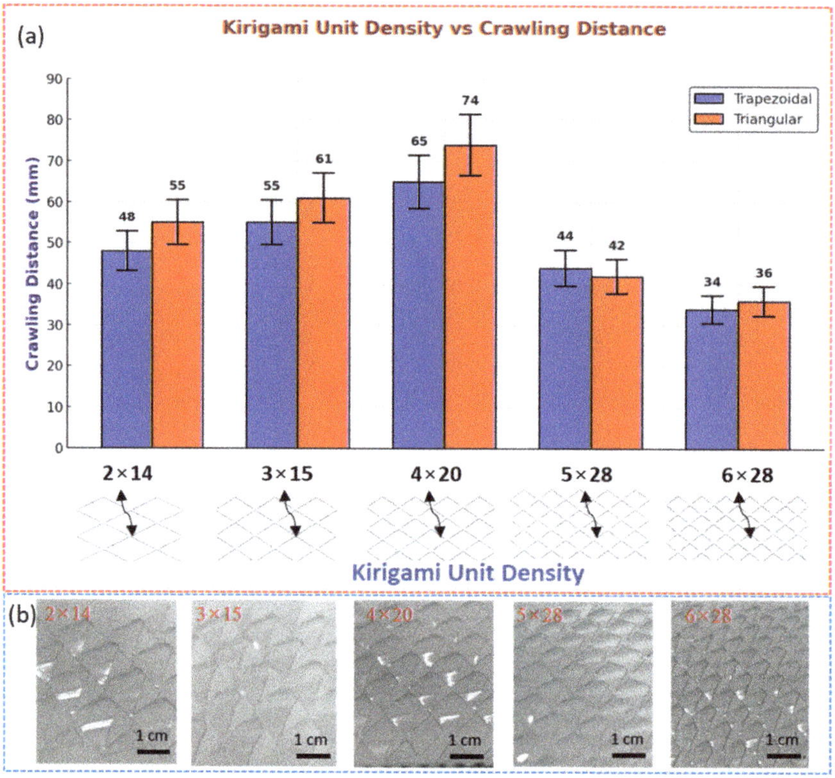

Figure 6. Influence of kirigami unit density on crawling performance and deformation characteristics. (**a**) Crawling distance comparison for triangular and trapezoidal kirigami skins across five different unit densities. The lower part is schematic illustrations of triangular kirigami patterns for each density, and at corresponding bars are experimental results for triangular and trapezoidal skins. (**b**) Experimental images of actuated triangular kirigami skins corresponding to the densities showing the pop-up deformation behavior.

To further validate these findings, we conducted frictional force analysis using a friction measuring setup (Section 2.2). As shown in Figure 7, the 4 × 20 configuration exhibited the highest stick–slip friction peaks (~3 N). In contrast, 5 × 28 and 6 × 28 configurations demonstrated lower peak values, confirming that excessive density reduces frictional effectiveness. Figure 6b includes side-by-side deformation images of kirigami samples at each density, clearly demonstrating the decreasing buckling amplitude and effective contact area. These results collectively confirm that an optimal configuration represents a balance between deformation, friction, and effective anchoring of kirigami units. These findings provide valuable insight into the bioinspired design of functional kirigami skins, particularly for use in confined environments where surface adaptability, frictional modulation, and efficient motion are essential.

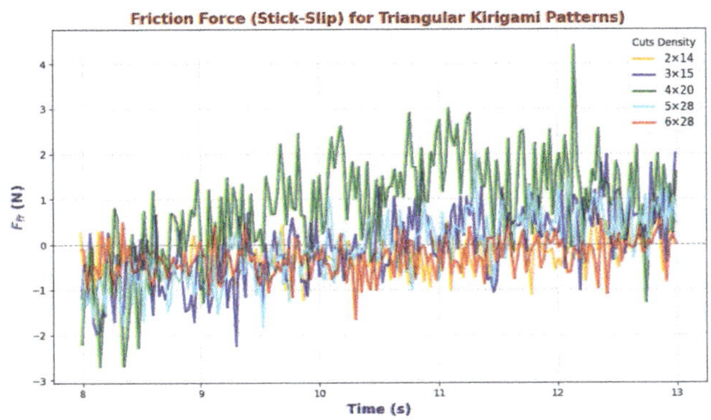

Figure 7. Friction force profiles for triangular kirigami patterns with measured varying unit densities.

3.4. *Stroke Length Variation in Microcontroller-Controlled Experiments*

3.4.1. Experimental Conditions

To further evaluate the locomotion performance of the soft robotic crawler, a series of experiments was conducted to analyze the stroke-to-crawl efficiency across varying actuation strategies. In this context, efficiency was defined as the ratio between the total crawling distance achieved by the robot and the total extension performed by the actuator over the duration of the test. The actuator extension was calculated as the product of the actuator's stroke length and the number of actuation cycles completed within the 30-second experiment window. The goal of this study was to understand how different combinations of stroke length and cycle timing influence the conversion of actuator motion into forward locomotion, thus complementing the density-based structural study detailed in Section 3.3. Four configurations were tested, all using triangular kirigami skin patterns fabricated from 100 μm thick PET sheets (with a Young's modulus of 4.33 GPa and Poisson's ratio of 0.4) and operated on a compliant polyurethane (PU) foam surface at a constant actuation speed of 15 mm/s. The tested stroke configurations were as follows: St10, which employed a 10 mm stroke and completed 15 cycles; St20, with a 20 mm stroke and 12 cycles; St30, which used a 30 mm stroke across 7 cycles; and St30h1s, which also used a 30 mm stroke but with an added 1-second hold time at the end of each stroke, completing 5 total cycles. These configurations resulted in total actuator extensions of 150 mm (St10), 240 mm (St20), 210 mm (St30), and 150 mm (St30h1s), respectively.

3.4.2. Performance Analysis

The results revealed significant variation in stroke-to-crawl efficiency. The St30h1s configuration achieved the highest crawling distance, recording 79 mm of forward displacement from just 150 mm of actuator motion—resulting in an efficiency of approximately 52.7%. This markedly outperformed the St30 configuration, which reached 67 mm of displacement from 210 mm of actuator extension (an efficiency of 31.0%). The lower stroke configurations, St20 and St10, exhibited reduced efficiencies of 19.2% and 17.3%, achieving crawling distances of 46 mm and 28 mm, respectively. These findings indicate that shorter stroke lengths, while allowing for more actuation cycles within a given time frame, result in lower overall displacement due to limited deformation amplitude and weaker surface engagement (Figure 8).

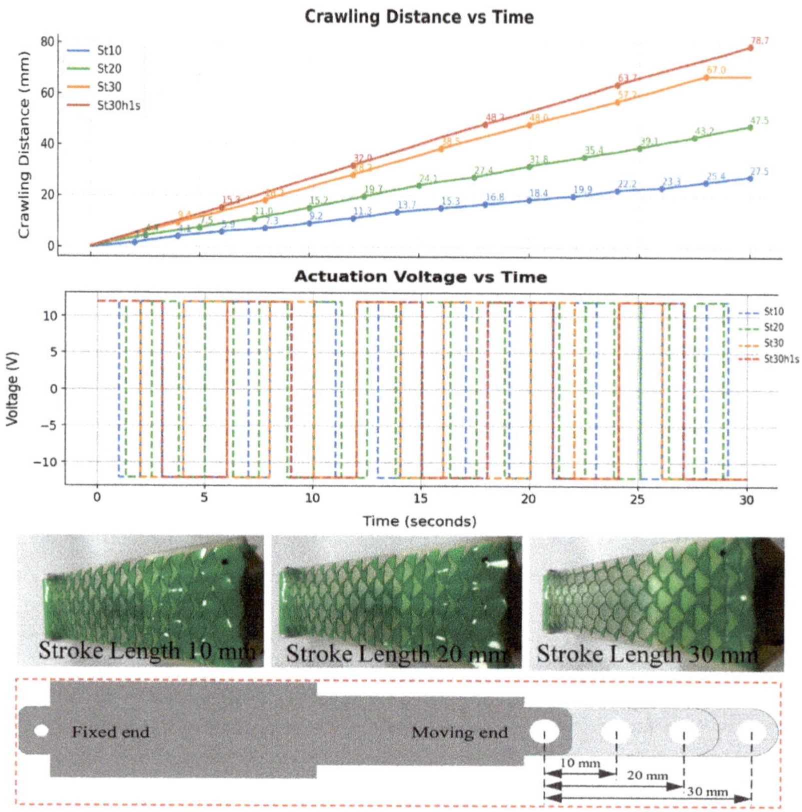

Figure 8. Crawling distance over time with voltage stroke cycles for different stroke cycle conditions. The top plot shows the crawling distance as solid lines, with corresponding actuation voltage cycles as dashed lines, over a 30-second period. The bottom insets illustrate the kirigami structure's deformation at each stroke length.

The superior performance of St30h1s is attributed to the strategic 1-second hold time incorporated at the end of each stroke. This pause enhances the robot's interaction with the substrate by allowing the kirigami skin to fully engage and conform to the surface, thus maximizing anisotropic friction during the transition between extension and retraction. The added hold time in St30h1s appears to exploit this frictional asymmetry more effectively, yielding a stride length of 15.8 mm per cycle, compared to 9.3 mm per cycle in St30. The data demonstrate that stroke length alone does not guarantee efficient locomotion. While St30 outperformed the shorter-stroke configurations, it still required greater actuator extension than St30h1s to approach similar displacement levels. This implies that timing and mechanical behavior during each cycle—particularly features like dwell time—can have a substantial impact on propulsion efficiency. By extending contact duration during the extension phase, the crawler benefits from improved anchoring and energy transfer to the ground, leading to more effective locomotion per unit of actuator work.

3.5. Optimization of Actuation Speed Parameters

3.5.1. Experimental Conditions

As discussed in the previous sections, not only does the applied strain on the kirigami skin play a critical role in deformation behavior, but dynamic actuation parameters also significantly influence the effectiveness of directional anchoring and overall locomotion performance. To investigate this, the present study evaluates the impact of actuation speed—specifically extension and retraction rates—on the crawling efficiency of a soft robotic system powered by a PM linear actuator. Experiments were conducted using a triangular kirigami pattern with a 4 × 20 cut density fabricated on a 100 µm PET sheet over a 51 mm × 140 mm contact area. The actuator was programmed to perform five extension–retraction cycles at a constant stroke length of 30 mm across four speed conditions defined by matched extension and retraction rates: 10/10, 15/15, 20/20, and 30/30 mm/s. These conditions were selected to evaluate how increasing the frequency of cyclic deformation affects the ability of the kirigami pop-ups to effectively anchor and re-engage with the substrate during locomotion. The goal was to determine whether higher actuation speeds reduce surface interaction time and compromise traction or whether they can maintain or even enhance performance through faster cycling.

3.5.2. Performance Analysis

Results shown in Figure 9 show that the crawling distances increased with actuation speed, peaking at the highest speed ratio. The 10/10 condition yielded 55 mm (5 cycles), averaging 11 mm per cycle. The 15/15 condition achieved 71 mm, averaging 14.2 mm per cycle. The 20/20 condition reached 91 mm, averaging 18.2 mm per cycle, while the 30/30 condition recorded 95 mm, averaging 19 mm per cycle. These results show that speed also impacts dynamic engagement rather than static friction anchoring force F_{prop}. This corresponds to a predicted per-cycle distance of ~19 mm (based on the model's scaling with F_{prop}), aligning with the observed 19 mm per cycle for 30/30.

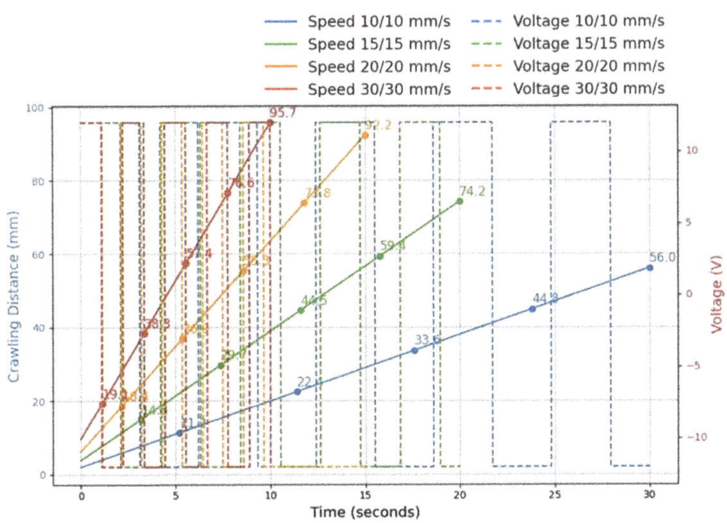

Figure 9. Crawling distance over time for different actuation speeds with corresponding actuation cycles shown as dashed lines.

These trends suggest that higher actuation speeds promote more complete deployment of the kirigami structures during each stroke, improving vertical pop-up height and surface engagement. In contrast, lower speeds result in slower, incomplete deployment, limiting anisotropic friction and reducing overall locomotion performance. To model this effect, a deployment factor was introduced—defined as the fraction of the actuator's stroke effectively converted into forward crawling. At 10 mm/s, the deployment factor is ~37%, consistent with the observed 11 mm stride. As speed increases, the factor improves to ~47%, 61%, and 63% at 15, 20, and 30 mm/s, respectively, aligning closely with experimental results. At higher speeds (20–30 mm/s), the actuator imparts greater momentum, allowing the kirigami units to fully buckle and anchor. This enhances anisotropic friction, resulting in more effective energy transfer and improved mechanical efficiency. These findings confirm that faster actuation cycles enable more complete kirigami engagement, thereby maximizing traction and stride length.

4. Autonomous Crawling of the Robot

Following a comprehensive experimental investigation of kirigami geometries, actuation speeds, and locomotion metrics, the triangular kirigami pattern with a 4×20 configuration ($l = 7$ mm, 100 μm PET sheet) was identified as the optimal design parameters for the autonomous soft robotic crawler. This configuration demonstrated superior crawling performance in experiments, leveraging enhanced pop-up deformation and pronounced anisotropic friction properties to generate effective surface anchoring and directional propulsion. The autonomous system builds upon microcontroller-driven experimentation and integrates a fully embedded control strategy. The prototype consists of an Arduino Nano microcontroller, a TB6612FNG motor driver (PCBSINO Technologies Limited, Shenzhen, China), and 12 V lithium battery (YG 12V, Zhejiang Yuyao Guotai Electronic Co., Ltd., Yuyao, China), with a 5 V regulated supply delivered via an LM7805 converter. The complete system is housed in a lightweight 150 g platform (Figure 10). To support untethered and repeatable operation, we developed a custom firmware loop on the Arduino Nano that manages actuator timing, direction control, and polarity switching. This autonomous loop enables cyclic crawling strokes with precise dwell and transition intervals, eliminating the need for external computation or command input. Actuation was programmed for a 30 mm stroke, using an extension speed of 30 mm/s, a 0.5-second dwell at full extension, and a 30 mm/s retraction speed. A polarity-switching control algorithm ensures smooth and repeatable actuation cycles, enabling the kirigami structure to engage the substrate during extension and retract cleanly without compromising the pop-up geometry. This compact and energy-efficient configuration leverages the low current draw of the PM actuator and the voltage regulation for stable performance under load. Moreover, the modularity of the electronics makes the system easily adaptable for future integration of sensors or control extensions [29]. Under autonomous operation, the crawler achieved a crawling speed of ~47 cm/min, with an average forward displacement of ~18 mm per 30 mm actuator cycle. This corresponds to a stride efficiency of approximately ~60%, confirming that nearly two-thirds of each actuator stroke contributed to productive locomotion (see Supplementary Video S1). These results validate the system's mechanical and structural optimization and reinforce the potential of kirigami-enhanced soft robotics for untethered and adaptable movement in constrained environments.

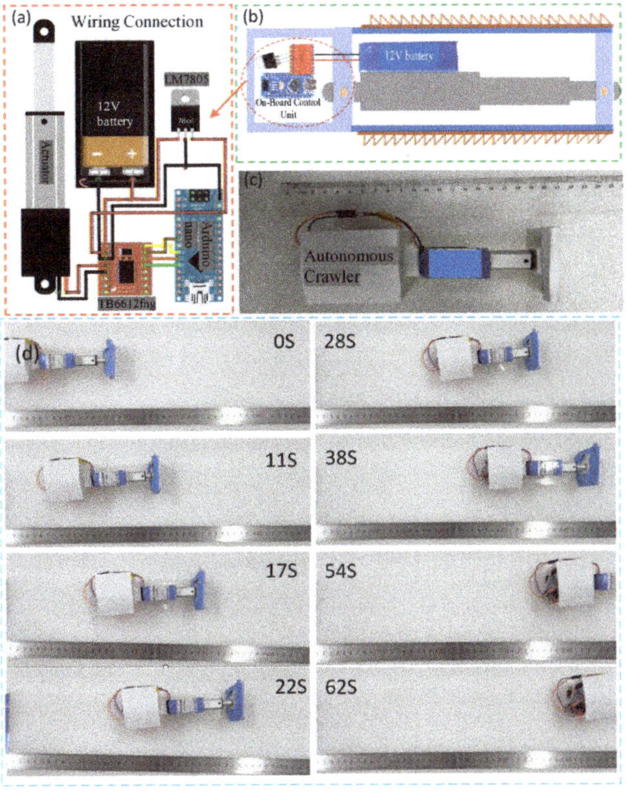

Figure 10. (**a–c**) Electrical integration and compact assembly of the autonomous kirigami crawler, featuring onboard actuation, battery, and control circuit for untethered operation. (**d**) Sequential locomotion snapshots over time showing directional crawling on a PU surface, demonstrating the effectiveness of the kirigami anchoring mechanism under autonomous actuation.

5. Conclusions

This study presents a bioinspired kirigami structure engineered to enhance the anchoring performance of a soft robotic system through systematic design and parameter optimization. Drawing inspiration from the anisotropic friction mechanisms observed in snake and lizard scales, we integrated patterned kirigami skins with a soft silicone body actuated by a compact linear actuator. The kirigami cuts comprising linear, triangular, trapezoidal, and hybrid geometries were evaluated for their ability to produce controlled buckling-induced deformation that translates into effective directional surface anchoring.

Through a comprehensive set of experiments, we analyzed how variations in kirigami geometry, sheet thickness, cut density, stroke length, and actuation speed impact anchoring efficiency, as quantified by crawling distance and stride performance. Our results show that triangular and hybrid designs offer significant improvements in propulsive force and surface engagement over baseline linear patterns. From a series of experiments, the triangular kirigami cuts on a 100 µm PET sheet with a 4 × 20 density were identified as the optimal configuration. Using these optimized design features and refined actuation parameters derived from experimentation, the kirigami-based crawler achieved a stride efficiency of approximately 63% and a maximum autonomous crawling speed of ~47 cm/min, confirming the configuration's effectiveness as a soft robotic anchoring mechanism.

Complementing the experimental results, we developed a theoretical model that combines buckling mechanics with frictional analysis to predict pop-up heights and directional friction coefficients. The strong agreement between predicted and observed performance metrics validates the model's utility for guiding future kirigami-based anchoring designs. Overall, this work demonstrates that kirigami-engineered skins, when coupled with efficient linear actuation and parameter tuning, provide a scalable and tunable approach to improving anchoring performance in soft robots. The proposed system offers a compact alternative to traditional pneumatic actuators and holds promise for applications requiring autonomous locomotion and high friction control in constrained or unstructured environments. Future work will focus on extending this platform through sensor integration and terrain-adaptive control to further enhance anchoring robustness and functional autonomy.

Supplementary Materials: The following supporting information can be downloaded at: https://www.mdpi.com/article/10.3390/app15147897/s1, Video S1: Demonstration of autonomous crawling performance of the kirigami-enhanced soft crawler.

Author Contributions: Conceptualization, methodology, validation, and software, M.N.K. and Z.S.; methodology, M.N.K.; formal analysis, M.N.K., Z.S. and M.Y.; investigation, Z.S., Y.H. and U.J.; data curation, M.N.K.; writing—original draft preparation, M.N.K.; writing—review and editing, M.N.K., Z.S. and U.J.; visualization, Y.H. and U.J.; supervision and funding acquisition, Z.S. All authors have read and agreed to the published version of the manuscript.

Funding: This work was supported by the National Key R&D Program of China (No.2022YFC3601100) and the National Natural Science Foundation of China (No. U19A20101).

Institutional Review Board Statement: Not applicable.

Informed Consent Statement: Not applicable.

Data Availability Statement: The data presented in this study are available on request from the corresponding author.

Conflicts of Interest: The authors declare no conflicts of interest.

References

1. Rus, D.; Tolley, M.T. Design, fabrication and control of soft robots. *Nature* **2015**, *521*, 467–475. [CrossRef] [PubMed]
2. Kim, S.; Laschi, C.; Trimmer, B. Soft robotics: A bioinspired evolution in robotics. *Trends Biotechnol.* **2013**, *31*, 287–294. [CrossRef] [PubMed]
3. Verl, A.; Albu-Schäffer, A.; Brock, O.; Raatz, A. *Soft Robotics: Transferring Theory to Application*; Springer: Berlin/Heidelberg, Germany, 2015; pp. 1–291. [CrossRef]
4. Hu, D.L.; Nirody, J.; Scott, T.; Shelley, M.J. The mechanics of slithering locomotion. *Proc. Natl. Acad. Sci. USA* **2009**, *106*, 10081–10085. [CrossRef] [PubMed]
5. Abdel-Aal, H.A.; El Mansori, M.; Zahouani, H. A comparative study of frictional response of shed snakeskin and human skin. *Wear* **2017**, *376–377*, 281–294. [CrossRef]
6. Gray, J. The Mechanism of Locomotion in Snakes. *J. Exp. Biol.* **1946**, *23*, 101–120. Available online: https://journals.biologists.com/jeb/article-abstract/23/2/101/12102 (accessed on 20 April 2025). [CrossRef] [PubMed]
7. Marvi, H.; Gong, C.; Gravish, N.; Astley, H.; Travers, M.; Hatton, R.L.; Mendelson, J.R.; Choset, H.; Hu, D.L.; Goldman, D.I. Sidewinding with minimal slip: Snake and robot ascent of sandy slopes. *Science* **2014**, *346*, 224–229. [CrossRef] [PubMed]
8. Parvaresh, A.; Seyidoğlu, B.; Rafsanjani, A. Metamaterial Soft Snake Robot for Rectilinear Locomotion. In Proceedings of the 2024 IEEE 7th International Conference on Soft Robotics (RoboSoft), San Diego, CA, USA, 14–17 April 2024; pp. 519–524. [CrossRef]
9. Lee, S.; Her, I.; Jung, W.; Hwang, Y. Snakeskin-Inspired 3D Printable Soft Robot Composed of Multi-Modular Vacuum-Powered Actuators. *Actuators* **2023**, *12*, 62. [CrossRef]

10. Rafsanjani, A.; Zhang, Y.; Liu, B.; Rubinstein, S.M.; Bertoldi, K. Kirigami skins make a simple soft actuator crawl. *Sci. Robot.* **2018**, *3*, eaar7555. [CrossRef] [PubMed]
11. Wang, C.; Puranam, V.R.; Misra, S.; Venkiteswaran, V.K. A Snake-Inspired Multi-Segmented Magnetic Soft Robot Towards Medical Applications. *IEEE Robot. Autom. Lett.* **2022**, *7*, 5795–5802. [CrossRef]
12. Grosso, B.F.; Mele, E.J. Bending Rules in Graphene Kirigami. *Phys. Rev. Lett.* **2015**, *115*, 195501. [CrossRef] [PubMed]
13. Rafsanjani, A.; Bertoldi, K. Buckling-Induced Kirigami. *Phys. Rev. Lett.* **2017**, *118*, 084301. [CrossRef] [PubMed]
14. Ganguly, S.; Margel, S. Fabrication and applications of magnetic polymer composites for soft robotics. *Micromachines* **2023**, *14*, 2173. [CrossRef] [PubMed]
15. Differing Ectoparasite Loads, Sexual Modes, and Abundances of Whiptail Lizards from Native and Non-Native Habitats. Available online: https://www.researchgate.net/publication/325102493_Differing_ectoparasite_loads_sexual_modes_and_abundances_of_whiptail_lizards_from_native_and_non-native_habitats (accessed on 23 April 2025).
16. Zheng, L.; Zhong, Y.; Gao, Y.; Li, J.; Zhang, Z.; Liu, Z.; Ren, L. Coupling Effect of Morphology and Mechanical Properties Contributes to the Tribological Behaviors of Snake Scales. *J. Bionic Eng.* **2018**, *15*, 481–493. [CrossRef]
17. Gans, C. *Biomechanics: An Approach to Vertebrate Biology*; University of Michigan Press: Ann Arbor, MI, USA, 1980.
18. Toro, A.; Abdel-Aal, H.; Zuluaga, E.; Cuervo, P.; Ballesteros, L.; Sánchez, J.; Rudas, J.; Isaza, C.; Misiolek, W. Influence of surface morphology and internal structure on the mechanical properties and tribological response of Boa Red Tail and Python Regius snake skin. *J. Mech. Behav. Biomed. Mater.* **2021**, *119*, 104497. [CrossRef] [PubMed]
19. Marvi, H.; Hu, D.L. Friction enhancement in concertina locomotion of snakes. *JRS Interface* **2012**, *9*, 3067–3680. [CrossRef] [PubMed]
20. Jayne, B.C. What Defines Different Modes of Snake Locomotion? *Integr. Comp. Biol.* **2020**, *60*, 156–170. [CrossRef] [PubMed]
21. Rafsanjani, A.; Jin, L.; Deng, B.; Bertoldi, K. Propagation of pop ups in kirigami shells. *Proc. Natl. Acad. Sci. USA* **2019**, *116*, 8200–8205. [CrossRef] [PubMed]
22. Mohammad, M.T.N.; Booth, C.N.; Hodgson, P.; Howlett, L.; Smith, P.J.; Sharkh, S.M. Design considerations and performance of a PM linear actuator in a radiation environment. In Proceedings of the 4th IET International Conference on Power Electronics, Machines and Drives (PEMD 2008), York, UK, 2–4 April 2008; pp. 757–761. [CrossRef]
23. Hasanzadeh, S.; Besharatifard, H. Analysis, Modeling, and Simulation of a PM Linear Actuator with Surface Motion Ability. In Proceedings of the 2023 14th Power Electronics, Drive Systems, and Technologies Conference (PEDSTC), Babol, Iran, 31 January–2 February 2023. [CrossRef]
24. Facebook. Available online: https://www.facebook.com/AfricanSnakebiteInstituteOfficial/photos/the-roughly-keeled-body-scales-of-a-puff-adder-bitis-arietans-arietans-snakes-sc/3730157157103687/?_rdr (accessed on 23 April 2025).
25. Snakes. In *Clinical Anatomy and Physiology of Exotic Species*; Saunders Ltd.: Philadelphia, PA, USA, 2005; pp. 77–93. [CrossRef]
26. Allam, A.A.; Abo-Eleneen, R.E.; Othman, S.I. Microstructure of scales in selected lizard species. *Saudi J. Biol. Sci.* **2019**, *26*, 129–136. [CrossRef] [PubMed]
27. El-Atab, N.; Mishra, R.B.; Al-Modaf, F.; Joharji, L.; Alsharif, A.A.; Alamoudi, H.; Diaz, M.; Qaiser, N.; Hussain, M.M. Soft Actuators for Soft Robotic Applications: A Review. *Adv. Intell. Syst.* **2020**, *2*, 2000128. [CrossRef]
28. 11.1: Compression Buckling of Thin Rectangular Plates—Engineering LibreTexts. Available online: https://eng.libretexts.org/Under_Construction/Aerospace_Structures_(Johnson)/11:_Buckling_of_columns_and_plates/11.01:_Compression_buckling_of_thin_rectangular_plates (accessed on 20 April 2025).
29. Qu, J.; Cui, G.; Li, Z.; Fang, S.; Zhang, X.; Liu, A.; Wang, X. Advanced flexible sensing technologies for soft robots. *Adv. Funct. Mater.* **2024**, *34*, 2401311. [CrossRef]

applied sciences

MDPI

Article

Towards a Digital Twin for Open-Frame Underwater Vehicles Using Evolutionary Algorithms

Félix Orjales [1], Julián Rodríguez-Cortegoso [1], Enrique Fernández-Pérez [1], Alejandro Romero [2] and Vicente Diaz-Casas [1,*]

[1] Integrated Group for Engineering Research, CITENI, Campus Industrial de Ferrol, Universidade da Coruña, 15403 Ferrol, Spain; felix.orjales@udc.es (F.O.); j.rcortegoso@udc.es (J.R.-C.); e.fperez@udc.es (E.F.-P.)

[2] Integrated Group for Engineering Research, CITIC, Universidade da Coruña, 15008 A Coruña, Spain; alejandro.romero.montero@udc.es

* Correspondence: vicente.diaz.casas@udc.es

Abstract: Hydrodynamic coefficients determine the behavior of all simulated underwater vehicles. Therefore, it is essential to precisely define their values when aiming to replicate a real vehicle. Generally established procedures for obtaining them tend to have limitations, especially in transient responses. To address these issues, this paper proposes a comprehensive methodology for obtaining the hydrodynamic coefficients of an underwater vehicle. The main novelty is the combination of empirical measurements as a first step and evolutionary algorithms as a final step for optimizing the coefficients. The proposed methodology is described and applied to a commercially available remotely operated vehicle (ROV) BlueROV2, followed by analyzing the results in detail and including several tests that compare it to the real vehicle to validate its adequacy.

Keywords: evolutionary algorithm; hydrodynamics; hydrodynamic coefficient; optimization; ROV; AUV; digital twin

1. Introduction

Most of the tasks performed by autonomous underwater vehicles (AUVs) have focused on long-range exploration, mostly large torpedo-type AUVs designed for visually exploring and mapping large areas that require very long ranges [1,2]. Inspection and maintenance tasks for parts of the installation that are above water are similar to those performed in onshore installations and are reasonably standardized. Underwater inspections, however, are in a different state, as there are still no procedures or systems in place to carry out these operations in a systematic and cost-effective way. The applications of AUVs in these tasks, such as undersea structure inspection, are relatively new; although, some developments have been made in the oil and gas sector, especially following oil or gas lines [3] and some basic ideas proposed for windfarms [4].

To address these tasks, an approach based on the coordination of swarms of small and medium-sized underwater vehicles could be useful. This approach involves deploying a group of AUVs with autonomous navigation and behavior capabilities, equipped with the appropriate sensor set and sensing strategies, to carry out the analysis of the state of offshore structures. These tasks involve the AUVs not operating in open waters, but in more confined spaces, and have hardly been addressed except for some work on cave exploration [5,6]. Each AUV in the team needs to accurately determine its position with respect to the structure and the rest of the team.

A digital twin of the AUV that allows integrating the measurements made in real time with the simulations of the dynamic behavior of the assembly is proposed in this work. This makes it possible to determine the current situation of the vehicle while simulating and evaluating the future outcome of the decisions to be made by the system. Although parts of the developments in other sectors are applicable to AUVs, the complexity of the environment in which it is located, the high nonlinearity of the environmental responses, and the high degree of interactions make this a very incipient area in the field of AUVs [7], especially when operating closely or in contact with large structures.

Understanding system behavior before real-world deployment is crucial for modern robotics. Digital twins (DTs) are especially valuable in marine robotics [8] where they support risk assessment, strategy optimization, and emergency planning [9] without exposing the physical system to danger. However, building high-fidelity virtual models and ensuring rapid data updates remain key challenges.

DT technology is expanding into cyber physical operation systems, allowing real-time interaction between physical robots and virtual environments. This integration is enhanced by virtual reality visualization and advanced physics engines, which simulate dynamic interactions—such as track-ground contact—with high accuracy. These innovations are validated through experimental comparisons and are paving the way for more stable, efficient, and intuitive underwater operations [8,10].

While DTs are widely applied in manufacturing and other sectors, their use in underwater environments is still maturing due to the complexity of marine modeling [11]. Nevertheless, ongoing research and technological integration are rapidly advancing the capabilities and impact of digital twins in marine robotics [12]. All these factors make the digital twin of AUVs an open problem today that requires not only a base platform but also specific models that allow representing the physical phenomena that take place during the activity of AUVs [7,13].

2. Related Work

Developing simulation tools has been a constant effort with the intention of offering a tool for all underwater works, as can be seen in reviews such as [8]. To model dynamic behavior, various aspects such as vehicle dynamics, propulsion, coefficients defining the model, and added masses need to be considered. In a more traditional simulation, all these models stay static while the vehicle evolves and changes; however, when working with a digital twin, both the vehicle and twin need to evolve together. This introduces a limitation into the evolution of the models as the evolution of the model also needs to be considered.

Regarding the dynamic model, it has been approached from multiple perspectives, covering a wide range of phenomena that can affect the dynamic behavior of underwater vehicles. Various strategies have been employed to validate the model, including numerical simulations, towing tank experiments, and wind tunnel tests. Early works, such as the dynamic behavior modeling of VideoRay Pro III, assume decoupled movements in the three axes [14]. In [15], low-cost sensors are used in real experiments to validate a simplified dynamic model, applied for AUV control in heading, pitch, and depth. Other cases, like [16], use finite element-based numerical examples to validate the dynamic modeling of umbilical cable effects on AUVs.

Dynamic models are also inferred from simulations using computational fluid dynamics (CFD) based on the meshing of the AUV's three-dimensional (3D) model [17]. Additionally, some studies focus on simulating the dynamic behavior of simple solids, as in [18], where models are tested using finite volume methods and validated in a wind tunnel, analyzing the effects of different edge shapes in detail, with results extrapolated to a real ROV.

In [19], a numerical model for a BlueROV2 is proposed, using the open source CFD software OpenFOAM to cover the four main freedom axes, and numerically simulated by a Reynolds-averaged Navier–Stokes (RANS) solver. The model is contrasted with hydrodynamic force and moment data measured in real trials, subjecting the ROV to various perturbations caused by different current and flow conditions.

Concerning the effects of vehicle propulsion on the dynamic model, different approaches have been considered, often motivated by vehicle control. In [20], the through-body thruster model of a torpedo-type C-SCOUT AUV is generated and numerically simulated, including forces due to hydrodynamic effects on control surfaces. Propeller shaft velocity and ambient flow velocity are used in [21] to estimate axial flow velocity, with the propulsion map divided into zones for analysis of the incoming angle of ambient flow through numerical simulation.

Some works address the propulsion model of underwater gliders propelled by buoyancy modulation. For instance, ref. [22] uses computational and semi-empirical methods to analyze the behavior of three different vehicles, ref. [23] examines thruster interactions, and ref. [24] investigates the influences of external currents on the propulsion model, conducting experiments with a BlueROV2 restrained by eight cables in different orientations relative to the incoming flow, and under different propeller operating conditions. With the development of open source simulators like Gazebo, many researchers started developing plugins and extensions to improve the accuracy of the underwater environment simulation such as the UUV Simulator [11], where the added mass, buoyancy, and other hydrodynamics are modeled using the Fossen equations [25].

As for the parameters affecting each model, various works focus on different aspects, and ref. [22] studies hydrodynamic parameters for the equations governing motion, while ref. [26] estimates hydrodynamic coefficients of a ROV using free decay pendulum motion. Real data are obtained in tests with a planar motion mechanism (PMM) for experimental evaluation of hydrodynamic coefficients in [23,27]. Furthermore, ref. [28] experimentally identifies hydrodynamic parameters for the LAURS open-frame AUV in one and three degrees of freedom, and ref. [29] uses experimental inertial measurement unit (IMU) values from an underwater vehicle to compare with results obtained through the genetic algorithm-based identification of hydrodynamic coefficients. Both refs. [30,31] evaluate hydrodynamic coefficients of an AUV, validating them through CFD simulations and experimental tests using a PMM. Furthermore, ref. [32] conducts towing tank and free-floating experiments to determine hydrodynamic coefficients of a ROV in 2 degrees of freedom.

The study of damping and added mass parameters usually accompanies the dynamic vehicle model. For instance, ref. [33] uses COMSOL Multiphysics to estimate added masses and damping parameters in the 6 degrees of freedom. However, comprehensive studies simultaneously addressing dynamic modeling, propulsion modeling, coefficients, and added masses are common, as seen in [26,34,35]. A comprehensive review is presented in [36].

In this paper, several important steps toward the development of a complete digital twin for an AUV are described. Specifically, Section 3 presents a comprehensive methodology for keeping the digital model up to date by combining information from traditional tests and measurements with the support of evolutionary algorithms to characterize the dynamic behavior of an AUV. An application of this methodology, using a BlueROV2 and a genetic algorithm, is presented in Section 4. The model is then validated through real-world tests, and the differences between model predictions and experimental results are analyzed and discussed in Section 5. Finally, Section 6 presents the conclusions drawn from the work.

3. Methodology Description

Over time, the underwater vehicles and their equipment are updated due to mission requirements, to improve their performance, or by obsolescence. The changes could affect the software, the equipment, and even the vehicle itself, and including more precise sensors, more capable actuators, or modifying other physical aspects can be quite common. All those changes have an impact on the real vehicle, changing its behavior. To avoid the divergence in results over time between the simulation and the reality, the digital twin needs to evolve in parallel with the real vehicle. This evolution requires the incorporation of some changes into the digital twin. This can be translated into the need to perform specific tests and measurements over the real vehicle to adequately evaluate the effects.

One of the key aspects that is affected in most cases is the vehicle's hydrodynamic behavior, particularly in the case of open-frame vehicles, whose geometry is highly dependent on the onboard equipment. In order to adjust the hydrodynamic model and accurately represent the real behavior, we propose a methodology based on evolutionary algorithms, which is described in this section. Figure 1 shows an overall diagram of the proposed process that is going to be explained in detail in the next subsections.

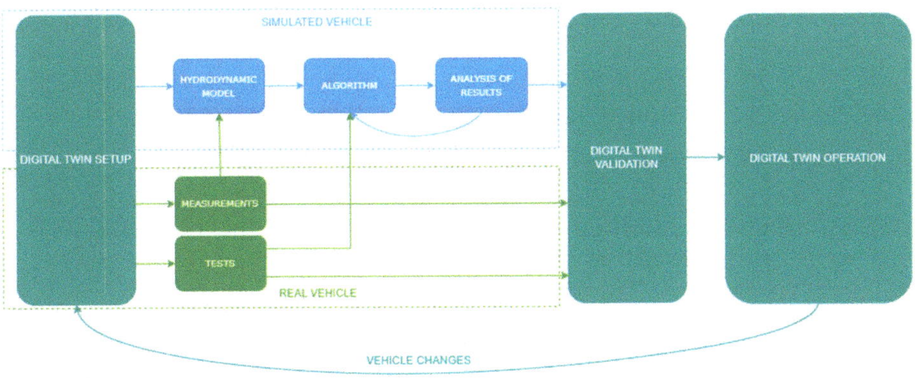

Figure 1. Methodology process diagram.

3.1. Digital Twin Setup

The first step involves creating an accurate representation by integrating various simulators and models and finding a way to have them communicate with the real vehicle. These models capture different aspects of the vehicle, such as its physical structure, sensors, actuators, electronics, and the environment in which it operates. The main purpose is to ensure that all necessary interactions and behaviors for the tasks that we want to perform are adequately represented, providing a solid foundation for subsequent simulations and analyses. Communication is also a crucial factor, as the simulated and real vehicle need to exchange information during the process.

3.2. Real Tests and Measurements

Once the setup is ready, the next step is to conduct different tests and measurements with the real vehicle. These tests are crucial for obtaining precise and reliable data that will serve as a reference for all the different steps adjusting the digital twin. Some of the tests are going to be direct and simple measurements, while in other cases, it may be necessary to perform estimations based on the results of the towing tank tests, position or orientation values over time, etc. To ensure the accuracy of the simulation model, it is necessary to cover a wide range of operating conditions, and the tests should be designed in such a way that they allow the acquisition of all the information needed to accurately model the real vehicle's behavior.

3.3. Hydrodynamic Model

Once the digital twin setup is established, the next crucial step is to estimate a hydrodynamic model, as a starting point for the optimization. There are many different methods to estimate the hydrodynamic coefficients; in some cases, they can be directly taken from similar vehicles in the literature [37]. In a more general approach, some methods are proposed in [38], including analytical methods, involving theoretical formulations and empirical correlations based on the vehicle geometry and operating conditions. Empirical methods involve conducting physical experiments, such as towing tank tests or free-decay experiments, to directly measure and observe the vehicle's response in controlled conditions. Numerical methods, particularly CFD, simulate fluid flow around the vehicle under different conditions, offering detailed insights into complex interactions between the vehicle's shape, propulsion systems, and surrounding fluid.

Both analytical and numerical methods can become quite complex, particularly when working with open-frame vehicles that present a complex geometry with holes and different shapes and a lot of equipment introducing disturbances (sensors, lights, cables, etc.). This leads to the empirical methods being the preferred in these applications as in [39], where the parameters of the dynamic model of the BlueROV2 have been determined and validated using both computer-aided design (CAD) software and physical experiments.

3.4. Optimization Algorithm

The previous estimation phase serves as a starting point, initializing the digital twin model for further refinement and validation through the proposed methodology. By combining the model with real data gathered from the vehicle, the hydrodynamic parameters of the model are optimized so the results of the simulations match the real data. This is a multi-objective problem, with a multitude of parameters to optimize, and a field where the use of a genetic algorithm (GA) has shown its flexibility and robustness.

GA is a stochastic, population-based optimization technique inspired by the process of natural selection and genetic evolution [40]. It operates by iteratively evolving a population of candidate solutions, with each solution represented as a set of parameter values. The optimization process aims to maximize a predefined fitness function, so that the simulated behavior closely matches empirical data from real-world trials.

3.5. Analysis of the Results

After the optimization process is finalized, the results are analyzed to detect some incoherencies or problems that could arise. The algorithm results are not filtered or directed in any way, only the upper and lower limits are considered for each parameter. The general idea with this step is to check for the feasibility of the results and to detect possible flaws. These flaws could be caused by some limitations on the model used, the fitness function considered, or the data-gathering process. However, they need to be adequately evaluated and addressed, if necessary, before the final validation tests.

3.6. Model Validation

The final step is to demonstrate the improvements achieved by the optimization process and the final state reached by the digital twin. This process involves conducting some tests with the real and simulated vehicles and a comparison of the results. This validation phase not only confirms the improvements made through optimization but also provides insights into future improvements and refinements of the digital twin over time.

3.7. Digital Twin Operation

After the digital twin is adjusted and mimics the real vehicle's behavior, it can be used as intended. During this stage, the digital twin allows the developers to test strategies using a realistic model of the real vehicle. This could be very useful for the development of novel coordination strategies, vehicle controllers, etc. This stage is only interrupted when there are significant differences between the real vehicle and the simulated twin. These changes are generally caused by modifications to the real vehicle required for a new task or project. When this happens, the process starts again, reusing everything from the previous iteration.

4. BlueROV2 Case

The vehicle selected for this work was a BlueROV2 from Blue Robotics Inc., an underwater open-frame ROV that has become very popular for its capabilities and affordability. More specifically, it was a BlueROV2 in its heavy configuration, with four horizontal propellers in a vectored position and the other four vertical propellers in each corner as shown in Figure 2.

Figure 2. BlueROV2 prepared for tests.

The BlueROV2 was powered by an on-board Li-Ion battery that provided up to 6 h of operation, and it was controlled by a Pixhawk autopilot. It was equipped with an IMU and other sensors and actuators like a pressure transmitter, leak-detection sensors, a low-light HD camera, lights, and a gripper. Onboard, there was also a Raspberry Pi 3 that was used in this project for high-level autonomous behaviors and to communicate through the tether or umbilical cord to a topside computer that was used as a supervision tool as described in our previous publication [41].

4.1. Digital Twin Setup

To test the desired autonomous behaviors for the inspection and maintenance of offshore structures using underwater vehicles without risking them or their equipment, a complete simulation environment was set up. This environment relied on the Gazebo [42] simulator with the Unmanned Underwater Vehicle Simulator (UUV Simulator) package [11] to simulate physical effects, such as the propellers, water currents, interactions with other objects, etc., but also included the control logic of the autopilot.

This approach allowed the same code to be executed on both real and simulated vehicles and to even use them simultaneously and seamlessly to test coordination strategies.

The mathematical modeling for the simulations was based on Fossen's [43] equations. For this study, the Society of Naval Architects and Marine Engineers (SNAME) notation [44] was used, and a summary is shown in Table 1.

Table 1. SNAME notation for marine vessels.

	Position and Angles	**Linear and Angular Velocities**	**Forces and Moments**
Movements	NED-frame	B-frame	B-frame
Surge	x	u	X
Sway	y	v	Y
Heave	z	w	Z
Roll	φ	p	K
Pitch	θ	q	M
Yaw	ψ	r	N

For a six-degree-of-freedom underwater vehicle, the matrix form of the equations of motion could then be expressed using the following:

M_{RB} as the rigid-body matrix.

$\dot{\nu}$ as the rate of change of the velocity ν with respect to time.

C_{RB} (ν) as the rigid-body Coriolis and centripetal matrix induced by M_{RB} due to the rotation of the body frame about the north, east, down (NED) world frame.

ν as the vehicle velocity vector.

M_A as the added mass matrix.

$\dot{v_w}$ as the rate of change of the velocity v_w with respect to time.

C_A (v_w) as the added mass Coriolis and centripetal matrix induced by M_A due to the rotation of the body frame about the NED world frame.

v_w as the relative velocity vector.

$D(v_w)$ as the damping matrix.

$g(\eta)$ as the restoring forces vector.

τ as a vector with all the external forces and moments actuating in the vehicle.

$$M_{RB}\dot{\nu} + C_{RB}(\nu)\nu + M_A\dot{v_w} + C_A(v_w)v_w + D(v_w)v_w + g(\eta) = \tau \quad (1)$$

The relative velocity vector v_w can be calculated using the vehicle velocity v and the current velocities in the fluid around the vehicle v_c and thus be expressed as:

$$v_w = \nu - v_c \quad (2)$$

A series of common assumptions when modeling underwater vehicles [45] could then be applied to simplify the equations and reduce the number of coefficients to be estimated:

- Assumption 1. The vehicle is assumed to be rigid, and 6 degrees of freedom (DOF) are considered.
- Assumption 2. The ROV is assumed symmetric around the front–back, port–starboard, and the top–bottom axes.
- Assumption 3. The body axes coincide with the main axes of inertia.
- Assumption 4. The origin of the b-frame is located at the center of mass of the vehicle.
- Assumption 5. The ocean current is modeled as a constant irrotational flow in the n-frame. Waves are neglected.

- Assumption 6. The movements for each DOF are assumed to be decoupled as they are performed at low speeds (less than 1 m/s).

The matrix expressions are detailed and explained in [43], but the most important ones are also included in this text to ease the lecture and interpretation. For symmetrical underwater vehicles during low-speed movements and with the gravity center on a vertical line from the coordinate origin (CO), the rigid body matrix M_{RB} and the added mass matrix M_A could be expressed using the following:

m as the mass of the vehicle in kg.
z_g as the vertical distance of the gravity center from the CO.
I_{XX}, I_{YY}, and I_{ZZ} as the moments of inertia for each axis.
$X_{\dot{u}}$, $Y_{\dot{v}}$, $Z_{\dot{w}}$, $K_{\dot{p}}$, $M_{\dot{q}}$, $N_{\dot{r}}$ as the added mass coefficients for each degree of freedom (DOF).

$$M_{RB} = \begin{Bmatrix} m & 0 & 0 & 0 & mz_g & 0 \\ 0 & m & 0 & -mz_g & 0 & 0 \\ 0 & 0 & m & 0 & 0 & 0 \\ 0 & -mz_g & 0 & I_{XX} & 0 & 0 \\ mz_g & 0 & 0 & 0 & I_{YY} & 0 \\ 0 & 0 & 0 & 0 & 0 & I_{ZZ} \end{Bmatrix} \quad (3)$$

$$M_A = -diag\left\{X_{\dot{u}}, Y_{\dot{v}}, Z_{\dot{w}}, K_{\dot{p}}, M_{\dot{q}}, N_{\dot{r}}\right\} \quad (4)$$

The Coriolis matrixes can be expressed as:

$$C_{RB}(\nu) = \begin{Bmatrix} 0 & 0 & 0 & 0 & mw & -mv \\ 0 & 0 & 0 & -mw & 0 & mu \\ 0 & 0 & 0 & mv & -mu & 0 \\ 0 & mw & -mv & 0 & I_z r & -I_y q \\ -mw & 0 & mu & -I_z r & 0 & I_x p \\ mv & -mu & 0 & I_y q & -I_x p & 0 \end{Bmatrix} \quad (5)$$

$$C_A(v_w) = \begin{Bmatrix} 0 & 0 & 0 & 0 & -Z_{\dot{w}}w & Y_{\dot{v}}v \\ 0 & 0 & 0 & Z_{\dot{w}}w & 0 & -X_{\dot{u}}u \\ 0 & 0 & 0 & -Y_{\dot{v}}v & X_{\dot{u}}u & 0 \\ 0 & -Z_{\dot{w}}w & Y_{\dot{v}}v & 0 & -N_{\dot{r}}r & M_{\dot{q}}q \\ Z_{\dot{w}}w & 0 & -X_{\dot{u}}u & N_{\dot{r}}r & 0 & -K_{\dot{p}}p \\ -Y_{\dot{v}}v & X_{\dot{u}}u & 0 & -M_{\dot{q}}q & K_{\dot{p}}p & 0 \end{Bmatrix} \quad (6)$$

The damping matrix is the combined expression of a lot of phenomena related to the interaction between the vehicle and the surrounding fluid, mostly dependent on the movement speed. It can be expressed as a second-degree polynomial of the movement speed for each axis, which implies two sets of parameters: one of them linearly coupled with the speed and one with the square of the speed.

$$D(v_w) = D_L + D_Q \times |v_w| \quad (7)$$

$$D_L = diag\{X_u, Y_v, Z_w, K_p, M_q, N_r\} \quad (8)$$

$$D_Q = diag\left\{X_{u|u|}, Y_{v|v|}, Z_{w|w|}, K_{p|p|}, M_{q|q|}, N_{r|r|}\right\} \quad (9)$$

The restoring forces vector $g(\eta)$ depends on η, the vector with the position and orientation of the CO relative to the navigation frame. For submersibles where the CO and center of gravity (CG) are in the same vertical line, they can be expressed using W and B as the weight and buoyancy of the vehicle, respectively, and z_b as the vertical distance of the buoyancy center from the CO as:

$$g(\eta) = \begin{Bmatrix} (W-B)\sin\theta \\ -(W-B)\cos\theta\sin\varphi \\ -(W-B)\cos\theta\cos\varphi \\ -z_b\, B\cos\theta\sin\varphi \\ -z_b\, B\sin\theta \\ 0 \end{Bmatrix} \quad (10)$$

Regarding the τ, the external forces and moments are primarily generated by the vehicle's thrusters and can be estimated from their position, orientation, and thrust. These values have been included in the model, and the thrust generated by each propeller is calculated in real time using the thruster curve.

$$\tau = \begin{Bmatrix} X \\ Y \\ Z \\ K \\ M \\ N \end{Bmatrix} \quad (11)$$

4.2. Real Tests

A series of real tests and measurements were performed to gather all the information required to model the vehicle and evaluate it in different conditions. The main characteristics and equipment used for the tests are described in each one of the subsections that follow.

4.2.1. Propeller Modeling

As the only part introducing energy over the system, the thrusters need to be measured and adequately characterized. There are individual characterizations of the BlueROV2 thrusters available [46] that could be used to generate an estimated curve for the vehicle based on the number of thrusters and their orientation. However, the propeller flow when mounted on the vehicle is partially interrupted by its structure, cables, etc.

To adequately measure the real thrust, a series of tests were performed using an AXI 9195 DELTA IP68 triaxle load cell to measure the empirical curve of the thrusters mounted on the vehicle. The tests consisted of thrust and torque measurements with the thrusters of each axis actuated for a few seconds with different turning speeds. The thruster rotation speed was controlled using an electronic speed controller (ESC) with a pulse width modulation (PWM) input signal.

Figure 3 shows the measured and estimated thrust for the x axis (X) at different PWM values of the input signal (I). The first curve was obtained directly from the measurements, while the estimated one was obtained from the individual curves and the geometry configuration. In this specific case, assuming the same response for the four horizontal thrusters and T as the individual thrust of each one, obtained from [46], we could express the estimated X as:

$$X = 4\,T\cos\frac{\pi}{4} \quad (12)$$

There are noticeable differences between the two curves, especially in the region with high PWM signals. The same phenomenon appears on the y axis thrust, with similar intensity. On the z axis, whose thruster flow is unimpeded, the estimated and the empirical curves are almost identical.

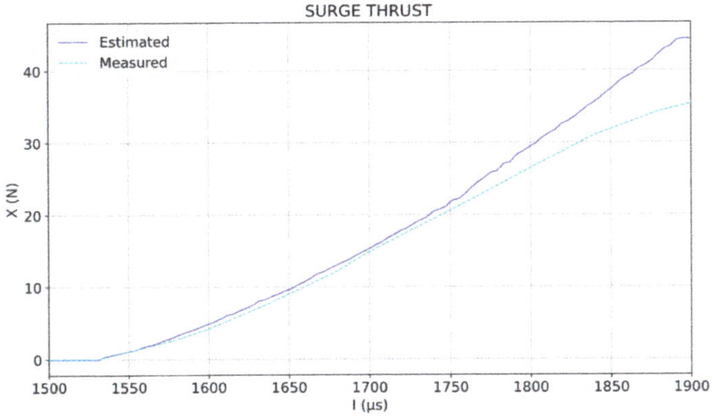

Figure 3. Estimated and measured thruster curve for surge movement.

4.2.2. Position and Orientation Tests

For the rotational movements (roll, pitch, and yaw), the tests consisted of movements in each axis at different thruster speeds. The motion was measured using the onboard IMU. For the linear motion tests (surge, sway, and heave), the ROV lacked a precise positioning system, so an underwater motion capture system from Qualisys was used. The system, shown in Figure 4, comprised a set of three Miqus underwater cameras that could detect and capture the motion of a body within a control volume up to 32 m^3. After the control volume was calibrated, five markers were placed asymmetrically on the vehicle frame, and a virtual representation of the rigid body was defined. Finally, the tests were performed by moving the vehicle in each axis at different speeds, and the positions were captured over time.

Figure 4. Setup for motion capture system.

4.3. Hydrodynamic Model

According to the measurements performed over the vehicle and some values from the literature, an initial hydrodynamic model was obtained. The results of all the estimations are shown in Table 2. A precision scale was used to measure the mass and buoyancy of the vehicle, while the CG was determined by balancing the vehicle in each axis and measuring the distance from the outside structure. To obtain an estimation of the vertical distance between the center of buoyancy (CB) and the CG, a series of weights were added in one side of the vehicle, generating a roll moment that was equilibrated by the moment caused by the CB. The resultant roll angles were measured using the integrated IMU of the vehicle.

Table 2. Initial estimations for hydrodynamic coefficients.

Parameters	Value	Units
m	13.17	kg
B	132.537	N
$CG\ (x_g,\ y_g,\ z_g)$	(0.0, 0.0, 0.0)	m
$CB(x_b, y_b, z_b)$	(0.0, 0.0, −0.024)	m
I_{XX}	0.344	Kg·m^2
I_{YY}	0.316	Kg·m^2
I_{ZZ}	0.389	Kg·m^2
$X_{\dot{u}}$	13.272	Kg
$Y_{\dot{v}}$	13.123	Kg
$Z_{\dot{w}}$	14.508	Kg
$K_{\dot{p}}$	0.207	Kg·m^2
$M_{\dot{q}}$	0.211	Kg·m^2
$N_{\dot{r}}$	0.109	Kg·m^2
X_u	−0.161	Kg/s
Y_v	−0.17	Kg/s
Z_w	−0.254	Kg/s
K_p	−0.349	Kg·m^2/s
M_q	−0.221	Kg·m^2/s
N_r	−0.141	Kg·m^2/s
$X_{u\|u\|}$	−33.346	Kg/m
$Y_{v\|v\|}$	−45.731	Kg/m
$Z_{w\|w\|}$	−72.668	Kg/m
$K_{p\|p\|}$	−0.356	Kg·m^2
$M_{q\|q\|}$	−0.461	Kg·m^2
$N_{r\|r\|}$	−0.471	Kg·m^2

Regarding the moments of inertia, they were estimated using CAD commercial software SOLIDWORKS 2023, and a model of the vehicle, in which the weights of the main components (battery, electronics, and payload) had been added into their actual position. For added mass and damping coefficients, estimations were made based on the vehicle geometry and similar vehicles from the literature.

4.4. Genetic Algorithm

In this subsection, we provide a detailed description of the GA employed for optimizing parameter values in our study. In our specific case, a genetic algorithm from the Python DEAP library (version 1.4.1) [47] was selected to optimize the values of the hydrodynamic model. The optimization process, represented in Figure 5, began with the initialization of a population of candidate solutions. Each individual in the population represented a candidate solution composed of a set of dynamic parameters (e.g., added mass, damping coefficients, and inertia), and the algorithm evolved the population over several generations to maximize a fitness function based on the similarity between the simulated and real trajectories.

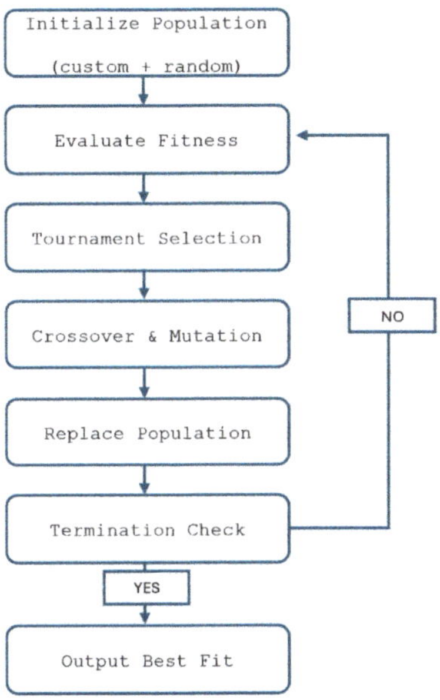

Figure 5. Block diagram illustrates the flow of the proposed genetic algorithm.

Each DOF was optimized individually, and the selected parameters for the optimization were different for each type of movement. For the linear movements, only three parameters were selected: the added mass, the linear damping, and the quadratic damping coefficients. For the rotational movements, the inertia was also included as a parameter, as its accuracy may not be good enough for the final modeling. Table 3 summarizes the parameters that were selected for each type of movement.

Table 3. Selected parameters for optimization.

Type of Movement		Inertia	Added Mass	Linear Damping	Quadratic Damping
Linear Movements	Surge	-	$X_{\dot{u}}$	X_u	$X_{u\|u\|}$
	Sway	-	$Y_{\dot{v}}$	Y_v	$Y_{v\|v\|}$
	Heave	-	$Z_{\dot{w}}$	Z_w	$Z_{w\|w\|}$
Rotational Movements	Roll	I_{xx}	$K_{\dot{p}}$	K_p	$K_{p\|p\|}$
	Pitch	I_{yy}	$M_{\dot{q}}$	M_q	$M_{q\|q\|}$
	Yaw	I_{zz}	$N_{\dot{r}}$	N_r	$N_{r\|r\|}$

The range for each parameter was carefully chosen from the initial values listed in Table 2, with some reasonable margin up and down, so the algorithm could find the optimal parameters without using very broad limits. This was made with two ideas in mind: the first one was reducing the optimization time, as it is related to the search space size; and the second one was to avoid configurations that could produce similar numerical results on the simulation but were unreal, like positive damping coefficients. The main idea was to optimize the parameters, adjusting them, without losing all the original information from the real tests.

i. Initialization

The first generation was partially initialized using specific parameter values obtained from experimental data. The rest of the individuals were randomly initialized within defined parameter ranges.

ii. Evaluation

After initialization, each individual (candidate solution) was evaluated using a custom fitness function, which calculated the sum of absolute differences between the simulated and real data across multiple tests and degrees of freedom (DOFs).

iii. Selection

The selection process determined which solutions will proceed to the next generation based on their fitness values. In our study, we employed tournament selection to choose parents for the next generation. This strategy involved randomly selecting a subset of solutions (tournament size = 5) from the population, and the one with the highest fitness was chosen. This method helped maintain diversity while applying selection pressure.

As previously stated, the fitness function evaluated the performance of each candidate solution in the population. It quantified how well the solution matched the desired behavior or characteristics of the system being optimized. In our study, the fitness function was defined as the sum of absolute differences between the ground truth data and the simulated data for each parameter over time.

The problem to solve is a multi-objective one. Even when considering each DOF as decoupled, there are multiple thruster speeds for each one. Specifically in this work, eight different thruster speeds were selected for each DOF. To find an optimal solution, the partial fitness functions for all these tests were combined, adding them into the final fitness function. A series of weights were applied in this process to improve the low-speed movements, as they were more interesting during the inspections.

iv. Crossover

In the crossover stage, pairs of parent solutions were selected from the population based on their fitness values. These parent solutions underwent a crossover operation, where portions of their parameter values were exchanged to produce offspring solutions. The crossover operation introduces diversity into the population by combining beneficial traits from different parent solutions. In our implementation, we used a two-point crossover strategy with a probability of 0.5 to generate offspring, where two random points were selected along the parameter vectors, and the segments between these points were swapped between parents.

v. Mutation

Following the crossover, the offspring solutions underwent a mutation operation, which introduced small random changes to their parameter values. Mutation helps to explore new regions of parameter space and prevents premature convergence to suboptimal solutions. We applied a Gaussian mutation operator, where random Gaussian noise was added to each parameter value with a mutation probability of 0.25, introducing variability to avoid local optima. The magnitude of the mutation was controlled by the standard deviation parameter.

vi. Replacement

The offspring replaced the previous population, and the optimization process continued for a predetermined number of generations or until a termination criterion was met. Common termination criteria include reaching a maximum number of generations, achieving a satisfactory level of fitness, or observing negligible improvement over successive generations. In our case, we established a maximum of 30 generations. Once the termination criterion was satisfied, the algorithm returned the fittest solution found during the optimization process. Table 4 summarizes all the relevant parameters and information related to the genetic algorithm.

Table 4. Summary of genetic algorithm settings.

Stage	Description	Parameter Values
Initialization	Custom initialization for the first generation; standard initialization for subsequent generations	Specific parameter values; random parameter values within predefined ranges Population size: 60
Fitness Function	Sum of absolute differences between ground truth and simulated data	-
Selection	Tournament selection strategy	Tournament size: 5
Crossover	Two-point crossover strategy	Cross rate: 0.5
Mutation	Gaussian mutation operator	Mutation rate: 0.25
Termination	Maximum number of generations	Generations: 30

It is important to highlight that the novelty in our implementation lay in the custom initialization of the first generation using experimentally informed parameters, which biased the search toward realistic solutions. Additionally, multiple DOFs were optimized independently, and the fitness function combined the outcomes of multiple tests with a weighted approach to emphasize low-speed movement accuracy, an aspect critical in underwater inspection scenarios.

4.5. Analysis of the Results

Once the algorithm reached a solution, it was analyzed to verify that it was adequate and realistic. Some of this work was completed by setting an adequate range that avoided completely unrealistic solutions. The final values obtained are shown in Table 5.

Table 5. Comparison between optimized and initial hydrodynamic coefficients for the BlueROV2.

Parameter	Initial Value	Optimized Value	Difference (%)
I_{XX}	0.344	0.371	7.85
I_{YY}	0.316	0.352	11.39
I_{ZZ}	0.389	0.426	9.51
$X_{\dot{u}}$	13.272	15.638	17.83
$Y_{\dot{v}}$	13.123	16.477	25.56
$Z_{\dot{w}}$	14.508	16.751	15.46
$K_{\dot{p}}$	0.207	0.157	−24.15

Table 5. *Cont.*

Parameter	Initial Value	Optimized Value	Difference (%)
$M_{\dot{q}}$	0.211	0.165	−21.8
$N_{\dot{r}}$	0.109	0.133	22.02
X_u	−0.161	−0.153	−4.97
Y_v	−0.17	−0.176	3.53
Z_w	−0.254	−0.242	−4.72
K_p	−0.349	−0.377	8.02
M_q	−0.221	−0.201	−9.05
N_r	−0.141	−0.127	−9.93
$X_{u\|u\|}$	−33.346	−34.972	4.88
$Y_{v\|v\|}$	−45.731	−43.118	−5.71
$Z_{w\|w\|}$	−72.668	−70.209	−3.38
$K_{p\|p\|}$	−0.356	−0.389	9.27
$M_{q\|q\|}$	−0.461	−0.427	−7.38
$N_{r\|r\|}$	−0.471	−0.425	−9.77

The inertia values are very close to the initial estimations using the CAD model within a range of around 10%; however, the real values are slightly larger. This difference may well be caused by small elements that are not included in the CAD model, such as screws, cables, or similar objects. The added mass for the linear movements seems to be relatively close to the initial estimations but less accurate than the previous one, with a ±25% variation. Initial estimations for these values were intended to be only a general guide, so these results seem adequate. Regarding the physical meaning, results for the linear movements seem correct as the sway and heave movements that have a bigger surface area and tend to move more water with them show higher values of added mass. The rotational movements also seem correct, with values a little bit smaller than the initial estimations, but with a very similar relation between them.

The linear and quadratic damping coefficients for the displacements show very small variations within a ±5 range, which also confirms that the initial estimations with the load cell tests were accurate. Regarding the rotations, both linear and quadratic damping coefficients for the rotations show a higher difference in percentage, with the quadratic ones much closer to the estimations. It is important to consider that the values of these coefficients are small, so even small deviations from the original values are translated into big percentages. These results confirm that the hydrodynamic properties of the vehicle were kept after the optimization process and should improve the model's behavior compared to the original one.

5. Digital Twin Model Validation

With the digital twin adjusted, a series of tests were performed to evaluate the effectiveness of the proposed solution. Figure 6 shows the position error (E_p) distribution for the linear validation tests, where the error values represent the difference between the real and simulated positions of the vehicle for each movement. Each group of boxplots corresponds to a specific linear motion: surge, sway, and heave. The tests consisted of six isolated movements along each axis, similar to those used during algorithm training, but with different amplitudes.

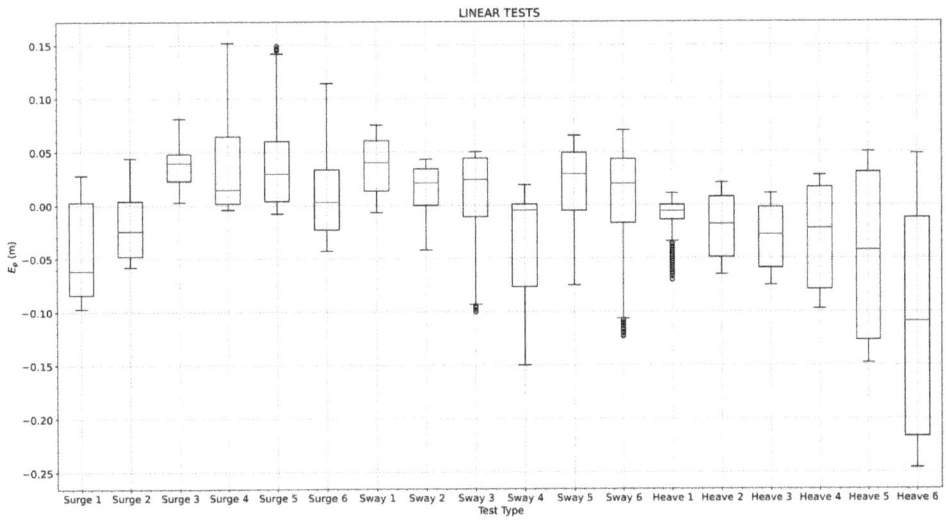

Figure 6. Error in the linear validation tests: 6 for surge, 6 for sway, and 6 for heave.

The amplitude ranges for the surge and sway motions were selected to ensure the robot exhibited representative behavior under typical operational conditions. In the case of heave motion, the range was limited by the depth of the towing tank. The results for the surge motion showed low error magnitudes, with most medians close to zero. This indicates that the simulated vehicle closely followed the trajectory of the real one. A similar trend was observed in the sway maneuvers, where the medians remained centered and the dispersion was moderate. The absence of a consistent trend in error magnitude with increasing amplitude suggests that the simulation accurately reproduced the vehicle's behavior across the range of velocities tested.

The heave tests exhibited different behaviors. While the initial tests (Heave 1 to 3) followed a similar trend to the other linear motions, the remaining tests (Heave 4 to 6) showed significantly larger errors and variability. This discrepancy may be attributed to disturbances generated by the interaction between the fluid and the bottom of the towing tank, which appeared more pronounced at faster heave displacements.

Figure 7 shows the orientation error (E_o) distribution for the rotational validation tests where the error values represent the difference between the real and simulated orientations of the vehicle for each rotational movement. Each group of boxplots corresponds to a specific rotation (roll, pitch, and yaw). For roll and pitch, six tests were carried out up to the stability limit for the robot in these degrees of freedom. Beyond that point, the vehicle became unstable and tended to capsize, making it difficult to obtain reliable or repeatable results. In contrast, yaw rotations were tested over a wider propeller speed range, provided that the motion remained controllable and representative of real operational conditions, as high speeds in these tests did not compromise the robot's stability.

The results for roll and pitch were consistent across all amplitudes, with low dispersion and median errors close to zero. This suggests a good match between the real and simulated responses for these two degrees of freedom. For yaw, the error values showed more variability, and the median was not centered around zero. This was expected, since the angular change in yaw was much larger than in roll or pitch, which naturally lead to a wider range of errors. Even so, the results followed a similar trend across the different test cases.

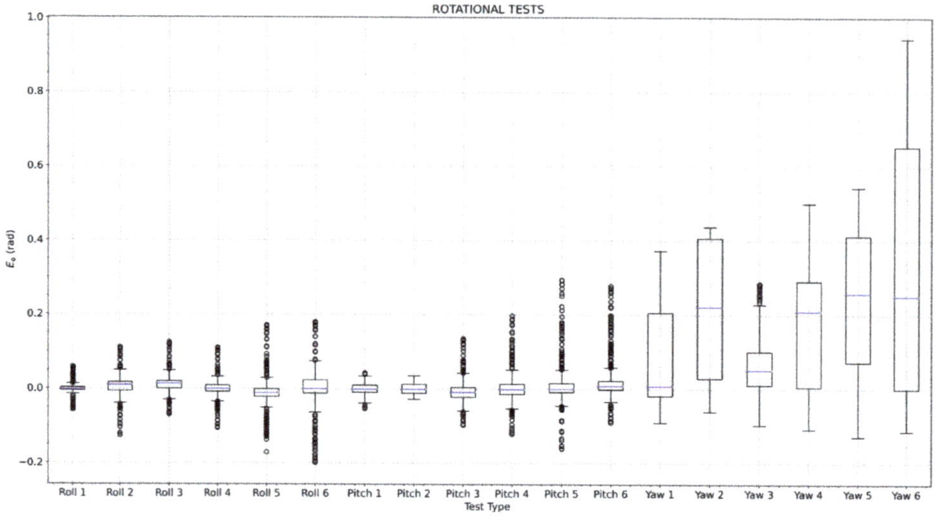

Figure 7. Error in the rotational validation tests: 6 for roll, 6 for pitch, and 6 for yaw.

A detailed comparison of the angle over time for one of the pitch tests is presented in Figure 8. As can be seen, both the actuated and free motion behavior of the model were very close to the real vehicle. To further quantify the agreement, the root mean square error (RMSE) was computed, yielding a value of 0.015 and a normalized RMSE of 3.5%. These results indicate a strong correlation between the simulated and experimental data, suggesting that the model and the parameter optimization process are effective.

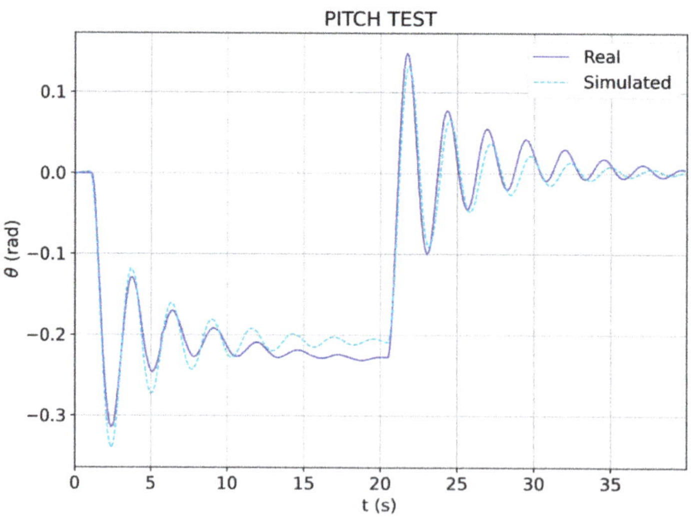

Figure 8. Pitch angle evolution over time in one of the pitch validation tests for the simulated and real vehicle.

To validate the decoupled motion assumption and assess the model's adequacy and generalization capability, a spiral ascent test was conducted. The results are illustrated in Figures 9 and 10, which respectively compare the heave displacement and yaw angle over time, for both the real system and the simulation model. Both the heave displacement and the yaw evolution during the combined maneuver exhibited a high level of agreement between the real and simulated data, particularly during the actuated phase of the motion. Minor deviations observed in the latter part of the test can be attributed to unmodeled external effects, such as disturbances introduced by the tether or the discarded coupling effects.

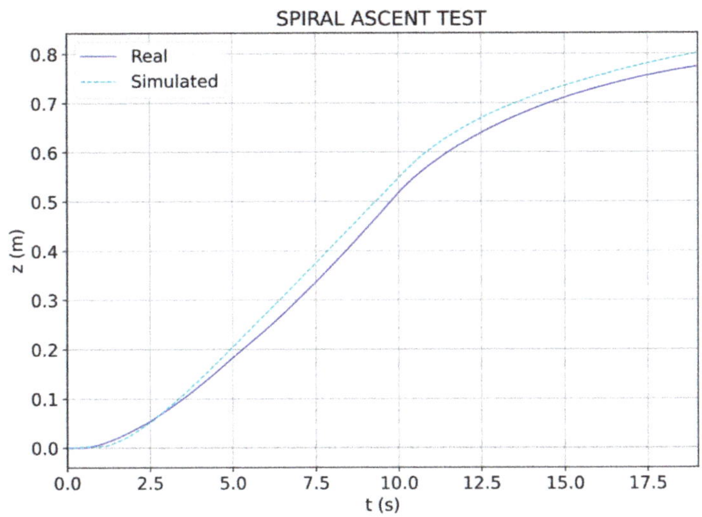

Figure 9. Heave displacement during spiral ascent test.

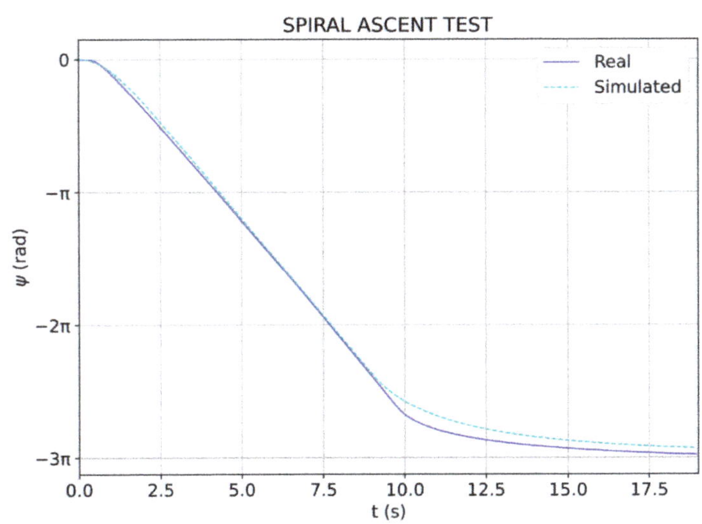

Figure 10. Yaw evolution during spiral ascent test.

The results, evidenced by the RMSE values of 0.03 (3.4%) for heave displacement and 0.17 (1.8%) for yaw angle, demonstrate that the model was capable of accurately capturing the translational and rotational dynamics independently, thus supporting the validity of the decoupling assumption made during model identification. Moreover, the ability to generalize to a complex, previously unseen trajectory involving simultaneous linear and rotational components highlights the model's robustness and predictive reliability in nontrivial operational scenarios.

6. Conclusions

In this paper, a methodology to improve the hydrodynamic coefficients of an underwater vehicle is proposed. The methodology is based on the combination of an initial estimation using traditional methods, the application of evolutionary algorithms to refine it, and a critical analysis of the results. This approach could lead to more accurate hydrodynamic coefficient estimations.

The results of the validation tests reveal significant improvements in the accuracy of the position estimations by the simulator when using the optimized model as compared to the original one. The behavior of the optimized model is much closer to the real vehicle, offering a viable alternative solution to the use of CFD software to improve these parameters.

The validation tests also revealed that there is still a reality gap, showing some differences between the real and simulated vehicle, particularly in the transient response. These variations may be caused by some limitations of the mathematical model, which does not consider some known phenomena such as the coupling between different movements or the effects of the umbilical cord. However, the same methodology could be applied to more complex models if required.

Author Contributions: Conceptualization, F.O. and V.D.-C.; Software, A.R.; Validation, A.R.; Formal analysis, E.F.-P.; Investigation, F.O., J.R.-C. and E.F.-P.; Writing—original draft, F.O., J.R.-C., E.F.-P. and A.R.; Writing—review & editing, V.D.-C.; Visualization, J.R.-C.; Supervision, F.O.; Project administration, V.D.-C. All authors have read and agreed to the published version of the manuscript.

Funding: This work has been partially funded by Grant TED2021-129847B-I00 by MCIN/AEI/10.13039/501100011033, by the "European Union NextGenerationEU/PRTR" and by the Xunta de Galicia ED431C 2021/39 program. This work has been partially funded by Grant PID2021-126220OB-I00 by MCIN/AEI/10.13039/501100011033 and by "ERDF A way of making Europe".

Institutional Review Board Statement: Not applicable.

Informed Consent Statement: Not applicable.

Data Availability Statement: The original contributions presented in this study are included in the article. Further inquiries can be directed to the corresponding author.

Conflicts of Interest: The authors declare no conflict of interest.

References

1. Lee, G.M.; Park, J.Y.; Kim, B.; Baek, H.; Park, S.; Shim, H.; Choi, G.; Kim, B.R.; Kang, H.G.; Jun, B.H.; et al. Development of an autonomous underwater vehicle ISiMI6000 for deep-sea observation. *Indian. J. Mar. Sci.* **2013**, *42*, 8.
2. Rumson, A.G. The application of fully unmanned robotic systems for inspection of subsea pipelines. *Ocean Eng.* **2021**, *235*, 109214. [CrossRef]
3. Albiez, J.; Joyeux, S.; Gaudig, C.; Hilljegerdes, J.; Kroffke, S.; Schoo, C.; Arnold, S.; Mimoso, G.; Alcantara, P.; Saback, R.; et al. FlatFish—A compact subsea-resident inspection AUV. In Proceedings of the OCEANS 2015—MTS/IEEE, Washington, DC, USA, 19–22 October 2015. [CrossRef]

4. Wright, M.; Gorma, W.; Luo, Y.; Post, M.; Xiao, Q.; Durrant, A. Multi-actuated AUV Body for Windfarm Inspection: Lessons from the Bio-inspired RoboFish Field Trials. In Proceedings of the 2020 IEEE/OES Autonomous Underwater Vehicles Symposium, AUV 2020, St. Johns, NL, Canada, 30 September–2 October 2020. [CrossRef]
5. Martins, A.; Almeida, J.; Almeida, C.; Matias, B.; Kapusniak, S.; Silva, E. EVA a hybrid ROV/AUV for underwater mining operations support. In Proceedings of the 2018 OCEANS—MTS/IEEE Kobe Techno-Oceans, Kobe, Japan, 28–31 May 2018. [CrossRef]
6. AMartins, A.; Almeida, J.; Almeida, C.; Pereira, R.; Sytnyk, D.; Soares, E.; Matias, B.; Pereira, T.; Silva, E. MARA—A modular underwater robot for confined spaces exploration. In Proceedings of the 2020 Global Oceans 2020: Singapore—U.S. Gulf Coast, Biloxi, MS, USA, 5–30 October 2020. [CrossRef]
7. Kutzke, D.T.; Carter, J.B.; Hartman, B.T. Subsystem selection for digital twin development: A case study on an unmanned underwater vehicle. *Ocean Eng.* **2021**, *223*, 108629. [CrossRef]
8. Matsebe, O.; Kumile, C.M.; Tlale, N.S. A Review of Virtual Simulators for Autonomous Underwater Vehicles (AUVs). *IFAC Proc. Vol.* **2008**, *41*, 31–37. [CrossRef]
9. Cho, H.-S.; Sohn, J.-H.; Han, J.-B.; Yeu, T.-K. Development of a Real-Time Track Solver for Digital Twin of the Underwater Tracked Vehicle. *Int. J. Precis. Eng. Manuf.-Green Technol.* **2025**, *12*, 1023–1036. [CrossRef]
10. Ciuccoli, N.; Screpanti, L.; Scaradozzi, D. Underwater Simulators Analysis for Digital Twinning. *IEEE Access* **2024**, *12*, 34306–34324. [CrossRef]
11. Manhães, M.M.M.; Scherer, S.A.; Voss, M.; Douat, L.R.; Rauschenbach, T. UUV Simulator: A Gazebo-based package for underwater intervention and multi-robot simulation. In Proceedings of the OCEANS 2016 MTS/IEEE Monterey, OCE 2016, Monterey, CA, USA, 19–23 September 2016. [CrossRef]
12. Scaradozzi, D.; Gioiello, F.; Ciuccoli, N.; Drap, P. A Digital Twin Infrastructure for NGC of ROV during Inspection. *Robotics* **2024**, *13*, 96. [CrossRef]
13. Adetunji, F.O.; Ellis, N.; Koskinopoulou, M.; Carlucho, I.; Petillot, Y.R. Digital Twins Below the Surface: Enhancing Underwater Teleoperation. In Proceedings of the OCEANS 2024, Singapore, 15–18 April 2024; pp. 1–8. [CrossRef]
14. Wang, W.; Clark, C.M. Modeling and Simulation of the VideoRay Pro III Underwater Vehicle. In Proceedings of the OCEANS 2006—Asia Pacific, Singapore, 16–19 May 2006; pp. 1–7. [CrossRef]
15. Valeriano-Medina, Y.; Martínez, A.; Hernández, L.; Sahli, H.; Rodríguez, Y.; Cañizares, J.R. Dynamic model for an autonomous underwater vehicle based on experimental data. *Math. Comput. Model. Dyn. Syst.* **2013**, *19*, 175–200. [CrossRef]
16. Eidsvik, O.A.; Schjølberg, I. Time Domain Modeling of ROV Umbilical using Beam Equations. *IFAC-PapersOnLine* **2016**, *49*, 452–457. [CrossRef]
17. Satria, D.; Wiryadinata, R.; A Esiswitoyo, D.P.; I Adji, M.; Rosyadi, I.; Listijorini, E. Hydrodynamic analysis of Remotely Operated Vehicle (ROV) Observation Class using CFD. *IOP Conf. Ser. Mater. Sci. Eng.* **2019**, *645*, 012014. [CrossRef]
18. Zarei, A.; Ashouri, A.; Hashemi, S.M.J.; Bushehri, S.A.S.F.; Izadpanah, E.; Amini, Y. Experimental and numerical study of hydrodynamic performance of remotely operated vehicle. *Ocean Eng.* **2020**, *212*, 107612. [CrossRef]
19. Li, Q.; Cao, Y.; Li, B.; Ingram, D.M.; Kiprakis, A. Numerical Modelling and Experimental Testing of the Hydrodynamic Characteristics for an Open-Frame Remotely Operated Vehicle. *J. Mar. Sci. Eng.* **2020**, *8*, 688. [CrossRef]
20. Evans, J.; Nahon, M. Dynamics modeling and performance evaluation of an autonomous underwater vehicle. *Ocean Eng.* **2004**, *31*, 1835–1858. [CrossRef]
21. Kim, J.; Chung, W.K. Accurate and practical thruster modeling for underwater vehicles. *Ocean Eng.* **2006**, *33*, 566–586. [CrossRef]
22. Geisbert, J.S. Hydrodynamic Modeling for Autonomous Underwater Vehicles Using Computational and Semi-Empirical Methods. Ph.D. Thesis, Virginia Tech, Blacksburg, VA, USA, 2007.
23. Avila, J.P.J.; Adamowski, J.C. Experimental evaluation of the hydrodynamic coefficients of a ROV through Morison's equation. *Ocean Eng.* **2011**, *38*, 2162–2170. [CrossRef]
24. Gabl, R.; Davey, T.; Cao, Y.; Li, Q.; Li, B.; Walker, K.L.; Giorgio-Serchi, F.; Aracri, S.; Kiprakis, A.; Stokes, A.A.; et al. Hydrodynamic loads on a restrained ROV under waves and current. *Ocean Eng.* **2021**, *234*, 109279. [CrossRef]
25. Fossen, T.I.; Fjellstad, O.-E. Nonlinear modelling of marine vehicles in 6 degrees of freedom. *Math. Model. Syst.* **1995**, *1*, 17–27. [CrossRef]
26. Lau, W.; Low, E.; Seet, G.; Chin, C. Estimation of the hydrodynamic coefficients of an ROV using free decay pendulum motion. *Eng. Lett.* **2008**, *16*, 326–331.
27. Avila, J.J.; Nishimoto, K.; Sampaio, C.M.; Adamowski, J.C. Experimental investigation of the hydrodynamic coefficients of a remotely operated vehicle using a planar motion mechanism. *J. Offshore Mech. Arct. Eng.* **2011**, *134*, 021601. [CrossRef]
28. Avila, J.P.J.; Donha, D.C.; Adamowski, J.C. Experimental model identification of open-frame underwater vehicles. *Ocean Eng.* **2013**, *60*, 81–94. [CrossRef]

29. Xie, S.; Li, Q.; Wu, P.; Luo, J.; Li, F.; Gu, J. Hydrodynamic coefficients identification and experimental investigation for an underwater vehicle. *Sens. Transducers* **2014**, *164*, 81–94.
30. Eidsvik, O.A. Identification of Hydrodynamic Parameters for Remotely Operated Vehicles. Master's Thesis, Norwegian University of Science and Technology, Trondheim, Norway, 2015.
31. Eidsvik, O.A.; Schjølberg, I. Determination of hydrodynamic parameters for remotely operated vehicles. In Proceedings of the International Conference on Offshore Mechanics and Arctic Engineering—OMAE, Busan, Republic of Korea, 19–24 June 2016. [CrossRef]
32. Lee, Y.; Lee, Y.; Chae, J.; Choi, H.T.; Yeu, T.K. Preliminary Experiments to Determine Hydrodynamic Coefficients of Remotely Operated Vehicle. In Proceedings of the OCEANS 2019—Marseille, Marseille, France, 17–20 June 2019. [CrossRef]
33. Hammoud, A.; Sahili, J.; Madi, M.; Maalouf, E. Design and dynamic modeling of ROVs: Estimating the damping and added mass parameters. *Ocean Eng.* **2021**, *239*, 109818. [CrossRef]
34. Rojas, J.; Eichhorn, M.; Baatar, G.; Matz, S.; Glotzbach, T. Parameter Identification and Optimization of an Oceanographic Monitoring Remotely Operated Vehicle. In Proceedings of the 2018 OCEANS—MTS/IEEE Kobe Techno-Oceans, Kobe, Japan, 28–31 May 2018. [CrossRef]
35. Soliman, M.A. Estimation of the ROV hydrodynamic coefficients using CFD. *Int. Undergrad. Res. Conf.* **2022**, *6*, 1–8.
36. Panda, J.P.; Mitra, A.; Warrior, H.V. A review on the hydrodynamic characteristics of autonomous underwater vehicles. *J. Eng. Marit. Environ.* **2020**, *235*, 15–29. [CrossRef]
37. Lack, S.; Rentzow, E.; Jeinsch, T. Experimental Parameter Identification for an open-frame ROV: Comparison of towing tank tests and open water self-propelled tests. *IFAC-Pap.* **2019**, *52*, 271–276. [CrossRef]
38. Busse, C.; Reich, J.E.; Renner, B.C. In Situ Damping Parameter Estimation for an Underwater Vehicle Using Onboard Sensors. In Proceedings of the 2022 IEEE/OES Autonomous Underwater Vehicles Symposium, Singapore, 19–21 September 2022. [CrossRef]
39. von Benzon, M.; Sørensen, F.F.; Uth, E.; Jouffroy, J.; Liniger, J.; Pedersen, S. An Open-Source Benchmark Simulator: Control of a BlueROV2 Underwater Robot. *J. Mar. Sci. Eng.* **2022**, *10*, 1898. [CrossRef]
40. Goldberg, D.E. Genetic and Evolutionary Algorithms Come of Age. *Commun. ACM* **1994**, *37*, 113–119. [CrossRef]
41. Rodriguez-Cortegoso, J.; Romero, A.; Orjales, F.; Deibe, A.; Diaz-Casas, V. Improving the Hydrodynamic Characterization of Autonomous Underwater Vehicles Through Deep Learning. In Proceedings of the The 33rd International Ocean and Polar Engineering Conference, Ottawa, ON, Canada, 19–23 June 2023; pp. 1764–1769.
42. Koenig, N.; Howard, A. Design and use paradigms for Gazebo, an Open-Source Multi-Robot Simulator. In Proceedings of the 2004 IEEE/RSJ International Conference on Intelligent Robots and Systems (IROS), Sendai, Japan, 28 September–2 October 2004. [CrossRef]
43. Fossen, T.I. *Handbook of Marine Craft Hydrodynamics and Motion Control*; Wiley: Hoboken, NJ, USA, 2021. [CrossRef]
44. Hydromechanics Subcommittee of the Technical and Research Committee of the Society of Naval Architects and Marine Engineers. *SNAME, Nomenclature for Treating the Motion of a Submerged Body Through a Fluid*; The Society of Naval Architects and Marine Engineers: New York, NY, USA, 1950; pp. 1–5.
45. Antonelli, G. *Underwater Robots*; Springer International Publishing: Cham, Switzerland, 2014; Volume 96. [CrossRef]
46. Lam, J.; Chen, A.; Bennett, A.; Triantafyllou, M. Propeller Characterization Testing of a Blue Robotics T200 Thruster. In Proceedings of the OCEANS 2023—Limerick, Limerick, Ireland, 5–8 June 2023; pp. 1–8. [CrossRef]
47. Fortin, F.A.; De Rainville, F.M.; Gardner, M.A.; Parizeau, M.; Gagńe, C. DEAP: Evolutionary algorithms made easy. *J. Mach. Learn. Res.* **2012**, *13*, 2171–2175.

applied sciences

MDPI

Article

A Machine Learning-Based Study on the Demand for Community Elderly Care Services in Central Urban Areas of Major Chinese Cities

Fang Wen [1,*], Zihao Liu [1], Bo Zhang [1], Yan Zhang [1], Ziqi Zhang [2] and Yuyang Zhang [1,*]

[1] Department of Architecture, School of Architecture and Art, North China University of Technology, Beijing 100144, China; liuzihao@mail.ncut.edu.cn (Z.L.); abaodoc@ncut.edu.cn (B.Z.); zhangyan@mail.ncut.edu.cn (Y.Z.)
[2] Department of Architecture, School of Design, Shanghai Jiao Tong University, Shanghai 200240, China; ziqizhang@sjtu.edu.cn
* Correspondence: fangwen@ncut.edu.cn (F.W.); yuyond@ncut.edu.cn (Y.Z.)

Abstract: China's population is aging rapidly, with a large proportion of elderly individuals "aging in place". In central areas of large cities, the amount of community and home-based elderly care services provided by the government and for-profit organizations are insufficient to meet the demands of these "aging in place" elderly. Taking the core area of Beijing as the spatial scope, this empirical study collects the demand on services of the main types of elderly residents in community and home-based dwelling through questionnaires (n = 242) and employs a mixed-methods approach for analysis. Descriptive statistics and exploratory factor analysis are used to determine the categories and levels of those demands, and machine learning methods (random forest regression model) are used to calculate the importance of various influencing factors (features of the elderly and subdistricts' built environment) on them. It is shown that elderly residents have a higher demand for psychological and physical condition maintenance services (mean = 3.40), and a lower demand for reconciliation and rights defense services (mean = 3.08). The results also show that the built environment factors are very important for the elderly on choosing demands, especially mean distance of CECSs (community elderly care stations) to downtown landmarks and main roads in subdistricts, and characteristics of CECS. The elderly's own features also have a relatively important impact, especially their living arrangements, caregivers, and occupations before retirement. This study applies machine learning techniques to sociological survey analysis, helping to understand the intensity of elderly people's demand for various community and home-based elderly care services. It provides a reference for the allocation of such service resources.

Keywords: machine learning; aging in place; demands for elderly care services; community elderly care facility; community elderly care station

1. Introduction

The increasing aging of the world's population has become a formidable challenge [1]. Global life expectancy is expected to reach 77.2 years by 2050 [2,3], and researchers predict that by 2080 there will be more people aged 65 and over than children under 18 [4]. The health and well-being of older people has become a very important issue [5], and there is already evidence that staying in a community or home environment for as long as possible has significant advantages over entering a nursing home [6–8]: the former promotes health and stability both physically and psychologically [9–11]. Therefore, it has become important to promote both active aging and aging in place in many parts of the world [12], which are seen as the best solution to cope with the rapidly growing number of older people [13]. Research on aging in place covers a wide range of areas such as housing, community, health, and social services [14], and community and home-based elderly care services have been found to be an important determinant of active aging of older people [15].

China has the largest elderly population in the world [16]. By the end of 2024, the population aged 60 and over amounts to 310.31 million individuals, accounting for 22.0% of the total population. Specifically, the population aged 65 and over is 220.23 million individuals, representing 15.6% of the total population [17]. The Chinese government advocates that more than 97% of the elderly should "age in place" [18,19], and this is also the preferred choice of the elderly themselves in China [20]. Aging in place not only slows down the physical and mental deterioration experienced by older people but also reduces the pressure on the construction of elderly care institutions, eases the strain on the allocation of care resources, and reduces public expenditures [21,22]. The role of community and home-based care services is crucial in supporting healthy and active aging in place, and studies have shown that using such services significantly improves older people's self-reported health [23], increases their life satisfaction [24,25], and promotes their physical and mental health [26].

However, the current elderly care services in China cannot fully meet the demands of those elderly who choose to age in place [27,28]. For example, in some communities there is a shortage of service staff, inadequate or misplaced services for different groups of older people, which leads to a decrease in satisfaction and acceptance of such services from older people [29–31]. In particular, for the care-dependent elderly, the low quality of services and the inability to achieve integrated care have greatly affected the sustainability of aging in place [32]. Therefore, clarification of Chinese older people's demands for all manner of elderly care services when aging in place is a crucial first step toward solving the problem [33].

The demands of older people for aging in place cover a wide range of categories, and there is great variability in different types of elderly groups. For example, health status has an impact on the demand for elderly care services [34]. Because younger and more vigorous older people are fully capable of self-management, they tend to focus on physical exercise, and cultural and recreational activities [24]. As physical function declines, the care demands of older people increase [35], requiring more "in place" support [36], such as Community Elderly Care Facilities (CECFs) that the Chinese elderly can use on a regular basis, or the purchase of a certain amount door-to-door services [37,38]. Adequate community and home-based care services are positively associated with the health status of older adults [23]. Of course, health problems and difficulties in accessing medical care increase with age, and most disabled older people find it difficult to avoid entering a nursing home [39,40]. Studies in China have shown that the main body of elderly living in community and home environments is dominated by healthy and assisted living, including a small proportion of nursing-cared elderly and even those who are completely unable to take care of themselves [41,42]. In recent years, due to the significant population growth of older people of advanced age; however, there has been an increase in the demand for professional care resources for those who want to live in a community environment [43]. In short, the care demands for those who age in place are diverse and vary from group to group [44], and this naturally has important implications for the planning and configuration of community elderly services and facilities.

The demands of older people who age in place are related to their socio-economic characteristics and family structure [40,45,46], such as income, education, and living arrangements [34,47,48]. Taking economic factors as an example, research indicates that individuals with poor economic conditions are more likely to fall into health poverty, with economically disadvantaged elderly experiencing higher rates of morbidity and mortality compared to groups with better economic status [49–51]. Coupled with an inability to afford nursing home care, poor elderly people must live in community and home environments for longer periods of time [52], and they require more affordable health and care resources [53,54]. Similarly, changes in the physical condition of older people influence their service demands, as do other factors such as age and gender [55,56]. There are also differences in the demands of specific elderly groups; for example, the demands of older people with cognitive disorders and disabilities are significantly different from those of healthy people [57]. Living arrangements and availability of caregivers are also determining factors for the demand for care services [58]. In addition, daily activities are associated with successful aging [59]. Such activities reflect the vitality and socialization patterns of older people and also indirectly influencing their demand for care services [60].

Differences in the built environment of residential areas also lead to differences in the demands of older people, especially demands on exclusive community care service resources. The condition of the built environment in residential areas, such as floor area ratio, housing price, road density, and accessibility to public transport, can indicate the degree of age-friendliness of a residential environment [61] and also influence the accessibility of elderly care services [62,63]. Adequate facilities in built environments can increase the life satisfaction of older people [24], which is conducive to their stable aging in place [64]. CECFs provide services exclusively for the elderly [38], and have been found to have an important link with the quality of life of elderly who age in place [38,65]. Studies in Lanzhou have found that the coverage of elderly care facilities and grassroots cultural facilities at all levels is seriously inadequate, however [66]. The distribution of CECFs of different types is also uneven, resulting in the demands of some older people not being met [67]. Furthermore, the study found that there is a general mismatch between function and demand, and insufficient usable space in the facilities in older communities [68].

Previous studies on the demand for elderly care services have focused on the distribution and location of resources, with the elderly population and entire urban areas as main indicators, which is spatially macroscopic. Analyses of the influencing factors of older people have mostly been explored in terms of their own socio-economic characteristics, and there is a lack of research on their built environments. This study therefore focuses on the community and home-dwelling elderly who live in urban core areas specifically as the target group in order to examine their demand for elderly care services more comprehensively. We explore how the multidimensional indicators of the elderly as individuals, the characteristics of the built environment, and community elderly care facilities impact the demand for aging in place at a subdistrict level. The aim of the study is thus to evaluate the demand for elderly care services on the basis of clear influencing factors, rather than from a macro-level allocation of resources. Also, the findings are conducive to promoting the allocation of elderly care services and provide further reference for the construction of community elderly care facilities, in order to better enhance the "aging in place" environment for the elderly people.

2. Materials and Methods

2.1. Study Area

The two central urban areas in Beijing, Dongcheng District, and Xicheng District, were selected as the study area (Figure 1). The aging problem in Beijing is relatively severe: by the end of 2021, the aging rate had already reached 20.2%, especially 27.0% in Dongcheng District and 26.9% in Xicheng District [69,70], which are the top two out of sixteen districts ranked by aging. Dongcheng District covers an area of 41.84 km^2, and Xicheng District covers 50.7 km^2, with high population densities and relatively well-developed public services in each. Due to the large number of elderly groups, the shortage of elderly care service is relatively pronounced. Subdistricts are administrative units under district level, and each subdistrict has unified management of residential areas, business areas, and public service facilities within its boundaries. Community elderly service facilities are also built and managed on a subdistrict basis. Driven by spatial convenience and administrative affiliation requirements, residents typically prioritize the use of service facilities within their own subdistrict. Dongcheng District currently has 17 subdistricts, and Xicheng District has 15 (Figure 2).

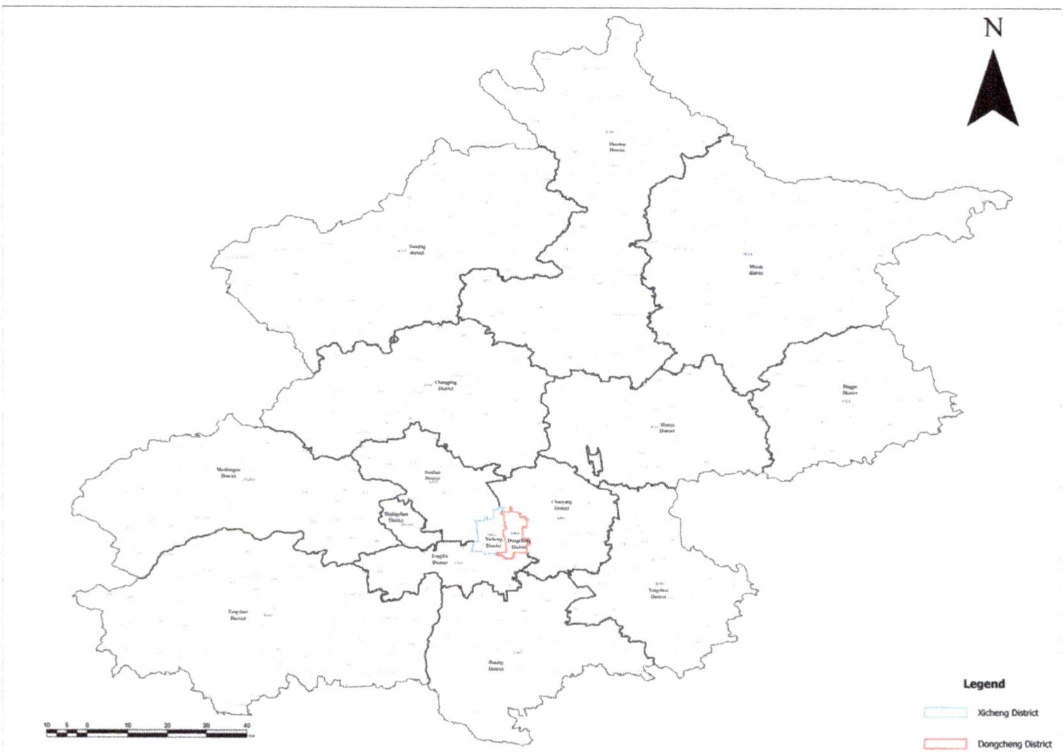

Figure 1. Dongcheng and Xicheng Districts in Beijing.

2.2. Research Framework

In the preliminary research stage, three parts of the study were prepared in advance. First, a large literature review was conducted to understand and summarize the demands of the aging in place elderly in urban area of large Chinese cities. Second, interviews were conducted with older people in several communities of Dongcheng and Xicheng regarding their demands for community and home-based care services. By comparing the demands with the first step, the interview results further clarify the similarities and differences in service demands among the elderly residents in the core areas of Beijing. Additionally, services provided by CECFs and their utilization in the two districts were investigated and summarized via on-site survey. All the above information was then combined into one dependent variable representing demand for care services. Through these preparatory efforts, the main types of service demands among the community-dwelling elderly population in the study area were ultimately identified, forming the dependent variable of this study. Future data on dependent variable (the elderly residents' attitudes toward these service demands) were collected via a questionnaire.

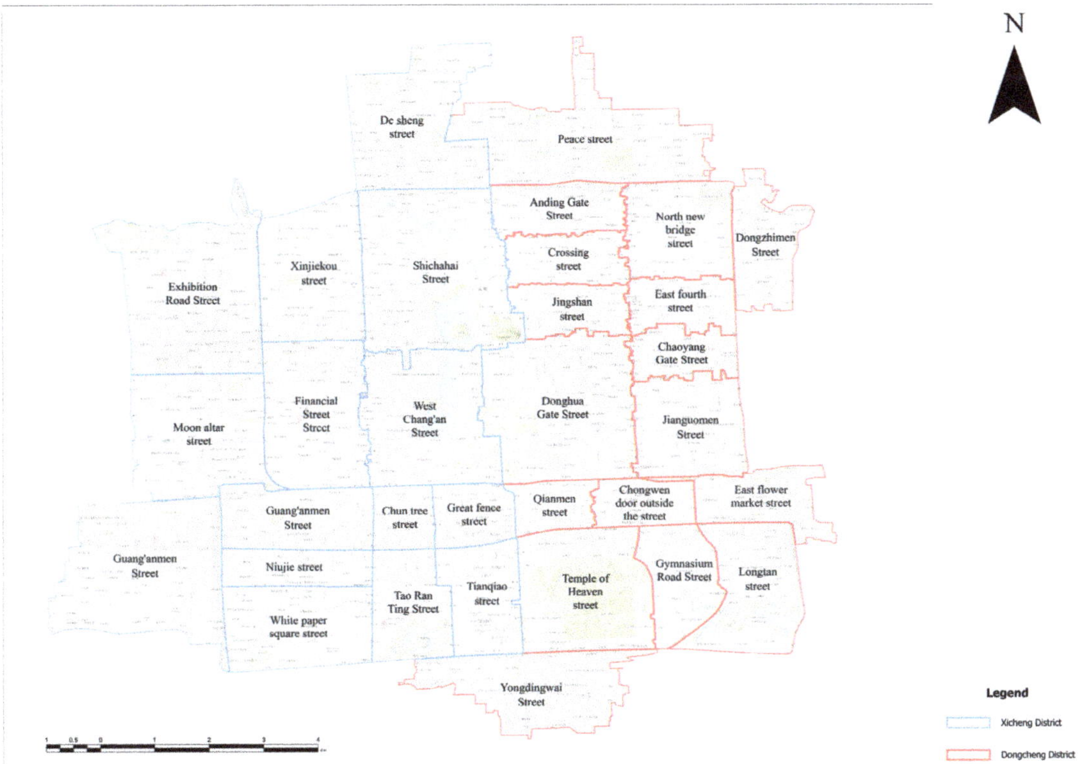

Figure 2. All subdistricts belonging to Dongcheng District and Xicheng District.

Subsequently, the independent variables of this study were determined, consisting mainly of two sets of data: the features of the elderly (including their attributes and daily activity characteristics) and the built environment of the subdistricts in which they reside. The data on the elderly residents were together collected through the same questionnaire used for the dependent variable. The built environment data were calculated using ArcGIS v.10.2 (Geographic Information System), encompassing general environmental spatial data and data on CECFs.

Prior to conducting the random forest regression analysis, basic analyses of the questionnaire data were performed using statistical methods. These included exploratory factor analysis to identify the main categories of various elderly care service demands and descriptive statistics to analyze the intensity of them. Thereafter, the focus was shifted to the analysis of machine learning results, the significant impacts of the elderly residents' personal attributes, and built environment characteristics factors (independent variables) on the elderly choosing various service demands (Figure 3).

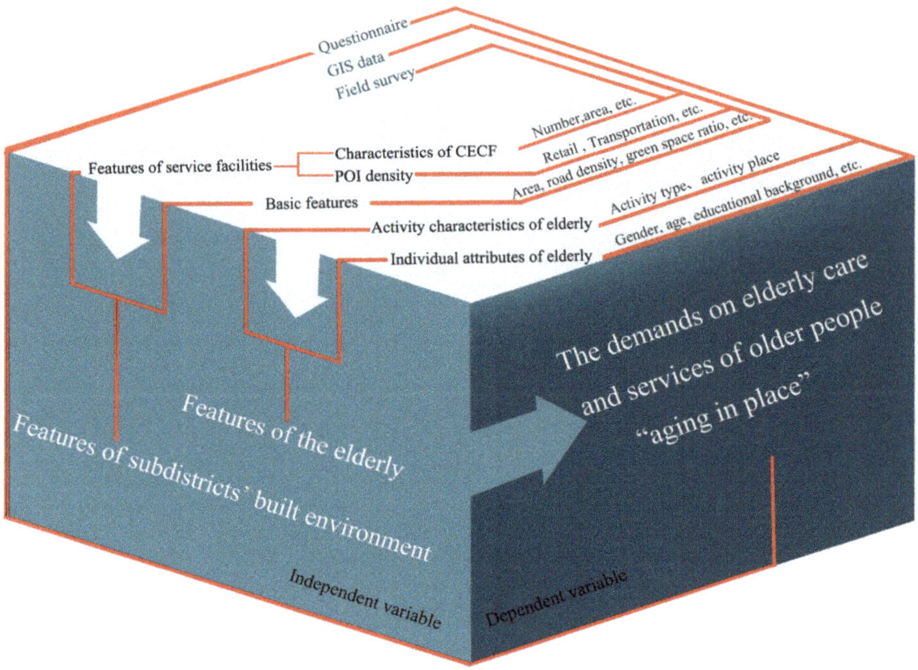

Figure 3. Flowchart of the machine learning part in the study.

2.3. Dependent Variables

Many previous studies have classified the demand for elderly care services into four main categories, though not always the same four. Some break down the demand into life care services (LCS), medical care services (MCS), spiritual and cultural services (SCS), and reconciliation and legal services (RLS), and some [24], while others break the demand into LCS, MCS, RLS, and spiritual and comfort services [34], and still others into life care, medical care, spiritual comfort, and economic assistance [71]. Under each of these categories, there are also more specific subcategories of services that cover various aspects such as older people's demand for diverse care services [72], their perceptions of the importance of door-to-door services [73], the demand for mobility assistance and travel companionship services [74], and the demand for leisure activities such as physical exercise [75]. The demand for spiritual care, health care knowledge, and neighborhood dispute resolution have also been studied [76].

Community elderly care station (CECS) is the collective term for CECFs in Beijing, and the six types of services required by the government in order to be a CECS, namely day care, emergency call service, dining, health guidance, psychological comfort, and cultural and recreational activities, illustrate the basic demand of community and home-dwelling older people. In the pre-survey stage of this study, we recorded and summarized the different types of services provided by CECSs of 11 randomly selected subdistricts and ultimately found that there are more services offered in CECSs beyond the required ones, such as education and legal assistance, which suggests that these services are also important to the community and home-dwelling elderly population.

In addition, semi-structured interviews were randomly conducted with elderly people in the communities of the study area to collect their views on their own demands. These collected demands were summarized and collated into categories, and the final summary is given in Table 1.

Table 1. Summary of 22 categories of the demands of older people.

Name	Contents
Meal preparation and delivery	a service that prepares meals and delivers them to a person's doorstep
Home cooking services	provides services for cooking up to three meals per day as needed
Dining spot for the elderly	dining spaces within walking distance in the community
Repairs and troubleshooting	regular or on-demand services to inspect and repair household appliances and check for safety hazards
Home cleaning	services that include cleaning the home, doing laundry, washing dishes, etc.
Shopping and grocery buying	services for shopping and purchasing groceries
Errands and payment services	includes picking up packages, handling various business payments, and calling for transportation services, etc.
Travel companionship	accompaniment to hospital, shopping, or other outings, and full-time companionship for outdoor activities
Routine care	haircuts, nail trimming, and assisted bathing provided at home
Home care	full-time or part-time home care services offering complete nursing care during the daytime
Emergency call service	effective calling for help to responsible departments or other emergency contacts
Community pharmacy services	dispensing medications at community health stations or hospitals
Rehabilitation and physiotherapy	various rehabilitation services, such as massage, strength training, traditional Chinese health care, etc.
Family doctor	providing regular consultations or follow-up services, including home visits for blood pressure checks, blood sugar testing, medical guidance, etc.
Health guidance	offering advice on senior nutrition, health care, common disease prevention, etc.
Cultural and recreational activities	engaging in cultural and entertainment activities, such as playing chess, singing, dancing, painting, etc.
Education	interest classes in calligraphy, painting, etc., as well as educational lectures
Physical exercise	conducting sports activities indoors or outdoors
Companionship and comfort	companionship, chatting, and emotional comfort services
Psychological counseling	regular or as-needed psychological consultations, emotional help, and interventional therapy
Legal assistance	providing legal advice or assistance as needed
Dispute resolution	assisting in resolving internal family disputes and neighborhood disputes

The above 22 categories were each taken to be dependent variables and were obtained from a questionnaire in the form of a 5-point Likert scale. A score of 1 represented "no need" and a score of 5 represented "extremely need".

2.4. Independent Variables

2.4.1. Older People's Own Attributes and Data

(1) Indicators of attributes

One set of independent variables is the socio-economic characteristics of the elderly, such as age, gender, living arrangements, monthly disposable income, etc., and the other set is the characteristics of their daily activities, including the type of daily activities and places (Table 2).

(2) Questionnaire design and data collection

Table 2. Indicators of the characteristics of the elderly themselves.

Indicators	Measurement Items
Gender	Male/Female
Age	50–59/60–69/70–79/80 and above
Educational background	junior high school or below/senior high school (including technical secondary school)/junior college/university
Occupation before retirement	agricultural production/government departments or public institutions/enterprise organization/self-employed business
Monthly disposable income	<3000/3000–5000/5000–8000/8000–10,000/≥10,000 CNY (RMB)
Living arrangements	alone/only with spouse/only with children/with spouse and children/only with grandchildren/with spouses and grandchildren/three generations/with other relatives/with a nanny or caregiver
Physical condition	independent/assisted/nursing care
Caregiver	spouse/children/other relatives/friends/hourly worker/live-in nanny or caregiver/self-care/none of the above
Daily activity	reading/walking/chatting/playing chess/singing/dancing/ball sports/Tai Chi/watching TV/playing cards/sunbathing/other
Activity places	outdoor spaces in community/parks or plazas nearby/senior activity rooms/community activity centers/other

Data for the above was obtained through a questionnaire and combined with the assessment of the needs in Table 1 (dependent variable) into one questionnaire. These were distributed during a two-week period in June 2023 in 24 subdistricts randomly selected from Dongcheng and Xicheng. Two periods, 8:00–10:00 a.m. and 4:00–5:30 p.m. in good weather were selected, when the temperature was favorable, and the elderly were presumed to be most active. The questionnaires were distributed near CECSs in each subdistrict, and the elderly who filled out the questionnaire have all been informed and have consented to the anonymous collection of their information. A total of 300 questionnaires were distributed, and the final number of valid questionnaires recovered was 242, covering 14 subdistricts in Dongcheng District and 10 subdistricts in Xicheng District (Figure 4). Because the study questionnaire does not collect sensitive personal information, does not involve commercial interests, and also does not cause harm to human subjects, ethics approval is not required for this type of study.

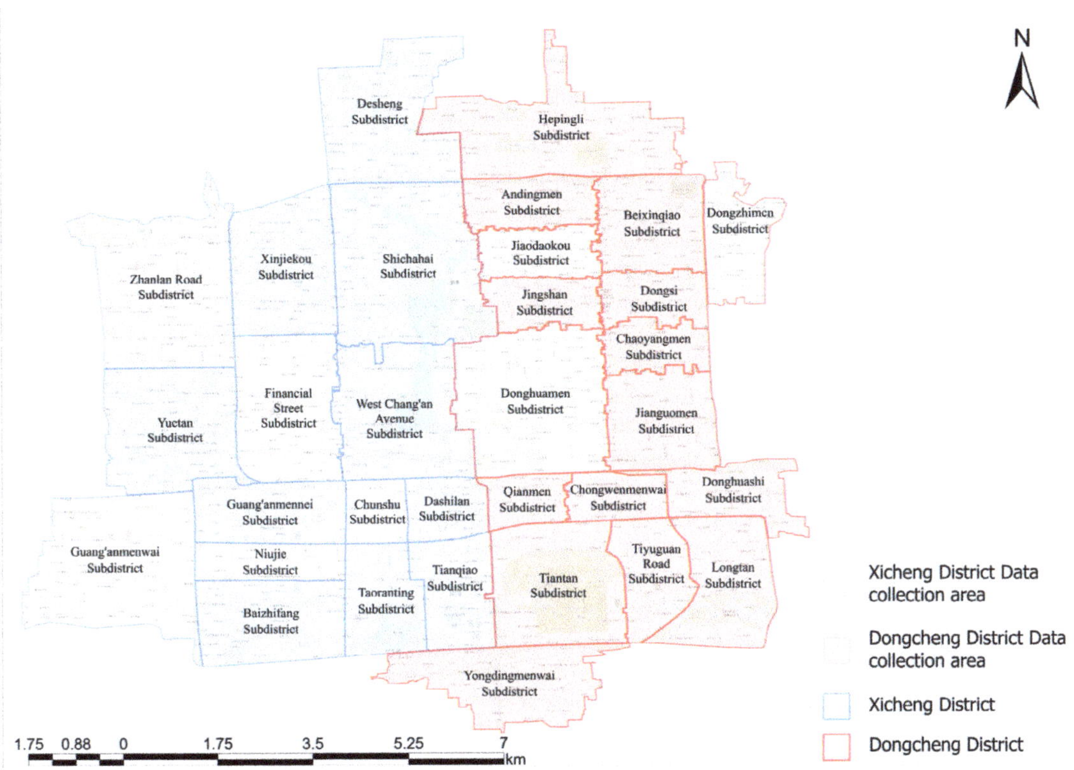

Figure 4. Twenty-four subdistricts covered by the questionnaire.

2.4.2. Built Environment Data

The built environment data also included two parts. One is the basic spatial indicators, such as road density and floor area ratio within each subdistrict, and the other is the characteristics of the CECSs such as their area, location, and accessibility. Therefore, ArcGIS 10.2 was used to prepare the built environment data on 24 subdistricts [77]. These were combined appropriately, and the built environment indicators were obtained, as shown in Table 3.

Table 3. Data indicators and sources for subdistricts.

Indicators	Calculation Method	Representation	Data Sources
Area of subdistrict	Total area of each subdistrict	Size of subdistrict	Amap
Road density	Sum of the lengths of all roads within each subdistrict and divided by the total subdistrict area	Accessibility	Amap, Map World
Floor area ratio	Sum of all building areas within each subdistrict and divided by the total subdistrict area	Degree of urbanization	Amap, OpenStreetMap

Table 3. *Cont.*

Indicators	Calculation Method	Representation	Data Sources
Number of bus stops	Sum of all bus stops within each subdistrict	Accessibility	Amap
Density of bus stops	Divide the number of bus stations by the total subdistrict area	Accessibility	Amap, OpenStreetMap
Housing price	Mean of all housing prices in all residential neighborhoods within the subdistrict	City location advantage	Amap
Green space ratio	Sum of all green areas and divide by the total subdistrict area	Residential environmental livability	Anjuke, Baidu
POI (Point of Interest) density	Number of different POIs (eating, shopping, science, education and culture, scenic spots, transport services, amenities, finance and insurance, sports and leisure, health care, etc.) divided by total road area.	Capacity of public services facilities	Amap

As mentioned above, in Beijing, CECS must have six types of service functions. They are typically well-managed and standardized so that CECFs [78] basically cover most of the needs of home-bound elderly people. Strictly speaking, CECFs are a class of public service facility, but because of the important role of community elderly service facilities in home care services for the elderly, the number and distribution characteristics of such facilities must be studied with separate indicators. In our study area, the spatial characteristics of CECS s in 24 subdistricts were collected (Table 4).

Table 4. Characteristic indicators of elderly service stations.

Indicators	Calculation Method	Representation	Data Sources
Number of CECSs	Sum of all CECSs in each subdistrict	Size of CECSs	Amap, On-site survey
Total area of CECSs	Sum of the areas of all CECSs within each subdistrict	Size of CECSs	On-site survey
Area ratio of all CECSs to subdistrict	Total area of all CECSs divided by area of the corresponding subdistrict	Coverage of CECSs	On-site survey, Amap
Mean distance from CECS to nearest main roads	Within each subdistrict, sum the distances from all CECSs to the nearest main roads, then divide by the number of CECSs	Accessibility	Amap, OpenStreetMap
Mean distance from CECS to nearest subway stations	Within each subdistrict, sum the distance from all CECSs to the nearest subway station, then divide by the number of CECSs	Accessibility	Amap, OpenStreetMap
Mean distance from CECS to downtown landmark	Within each subdistrict, sum the distances from all CECSs to Tiananmen Square, then divide by the number of CECSs	Accessibility	Amap, OpenStreetMap

2.4.3. Calculation Process

Exploratory factor analysis was used to examine the relationships between all of the above data points after calculating a KMO greater than 0.6, and $p < 0.05$ for Bartlett's sphericity test. A table of explained variance explained was produced, which was rotated using the maximum variance rotation method (varimax) to obtain a table of rotated factor loading coefficients that grouped the 22 categories of needs into five main areas.

In addition, machine learning techniques were also used to analyze the data [79–81]. First, the questionnaire data, the subdistrict variable data, and the POI data were red using the Python 3.12 Pandas library and merged based on the subdistrict name where the respondents reside. Next, feature columns were selected and StandardScaler was employed to standardize these features. Following preprocessing, the data were divided into training and test sets in an 8:2 ratio using the sklearn library. A random forest classification model was then trained with parameters n_estimators = 50 and random_state = 42. Upon completion of the model training, its predictive accuracy on the test set was calculated. Finally, the SHAP package was employed to generate SHAP bar plots, which elucidate the significance each variable within the random forest model.

3. Results

3.1. *The Intensity of Elderly People's Demand for Various Services and Analysis of Differences*

The 242 questionnaires were tested for reliability and validity, and yielded Cronbach's alpha = 0.915 and KMO = 0.905, indicating that the reliability and validity of the questionnaires were both good. Then, the 22 categories of demand were condensed into five major aspects by using exploratory factor analysis: living services, nursing care services, health care services, psychological and physical condition maintenance, and reconciliation and rights defense services (Table 5).

Table 5. Factor loading coefficients after rotation.

Summary of Demands in Five Main Areas	Name of Demands	Factor Loading					Communality
		Factor 1	Factor 2	Factor 3	Factor 4	Factor 5	
Living services	Meal preparation and delivery	**0.786**	0.212	0.155	0.269	0.097	0.768
	Home cooking services	**0.765**	0.012	0.105	0.150	0.177	0.650
	Dining spot for the elderly	**0.715**	0.232	−0.010	0.136	−0.106	0.595
	Repair and troubleshooting	**0.682**	0.176	0.209	0.123	0.058	0.558
	Home cleaning	**0.721**	0.123	0.169	0.182	0.027	0.597
	Shopping and grocery buying	**0.656**	0.113	0.160	0.102	0.173	0.509
	Errands and payment service	**0.669**	0.156	0.201	0.141	0.125	0.547
	Travel companionship	**0.742**	0.107	0.232	−0.029	0.118	0.630
Nursing care services	Routine care	0.180	0.110	0.109	**0.837**	0.087	0.765
	Home care	0.209	0.114	0.139	**0.774**	0.130	0.692
	Emergency call service	0.255	0.166	0.249	**0.617**	0.200	0.575
Health care services	Community pharmacy services	0.182	0.205	**0.732**	0.153	0.033	0.634
	Rehabilitation and physiotherapy	0.181	0.090	**0.779**	0.080	0.080	0.661
	Family doctor	0.216	0.100	**0.702**	0.100	0.140	0.579
	Health guidance	0.181	0.223	**0.748**	0.159	0.054	0.670

Table 5. *Cont.*

Summary of Demands in Five Main Areas	Name of Demands	Factor Loading					Communality
		Factor 1	Factor 2	Factor 3	Factor 4	Factor 5	
Psychological and physical condition maintenance	Cultural and recreational activities	0.133	**0.791**	0.125	0.147	0.008	0.681
	Education	0.170	**0.720**	0.206	0.057	-0.008	0.593
	Physical exercise	0.221	**0.695**	0.177	0.046	0.099	0.576
	Companionship and comfort	0.091	**0.720**	0.013	0.082	0.318	0.634
	Psychological counseling	0.284	**0.508**	0.304	0.278	0.157	0.533
Reconciliation and rights defense services	Legal assistance	0.144	0.180	0.170	0.126	**0.830**	0.786
	Dispute resolution	0.179	0.128	0.081	0.213	**0.807**	0.751

Rotation method: varimax maximum variance.

An analysis of the frequency means of the options of the five main types of demands showed that the most frequently chosen option was "3", followed by "4", "2", and "5" points were chosen more or less the same quantity, and "1" (no need) was chosen by the fewest people. This suggests that there is a relatively high level of demand for elderly services, with only a minority of cases indicating no demand (Figure 5). Among the five main demands, the highest demand was for psychological and physical condition maintenance (mean = 3.40) and the lowest demand was for reconciliation and rights defense services (mean = 3.08). Among the 22 subcategories, the ones with the three highest levels of demand were cultural and recreational activities, repair and troubleshooting, and emergency call service, which belong to the three categories of psychological and physical condition maintenance, living services and nursing care services, respectively. These popular services were followed by physical exercise, companionship and comfort, community pharmacy services, education, health guidance, rehabilitation and physiotherapy, and routine care (mean > 3.3), which belong to the three categories of psychological and physical condition maintenance, nursing care service and health care service, respectively. Overall, older people had a low demand for living services, although the lowest demand was for legal assistance and dispute resolution, the only two types in the category of reconciliation and rights defense services.

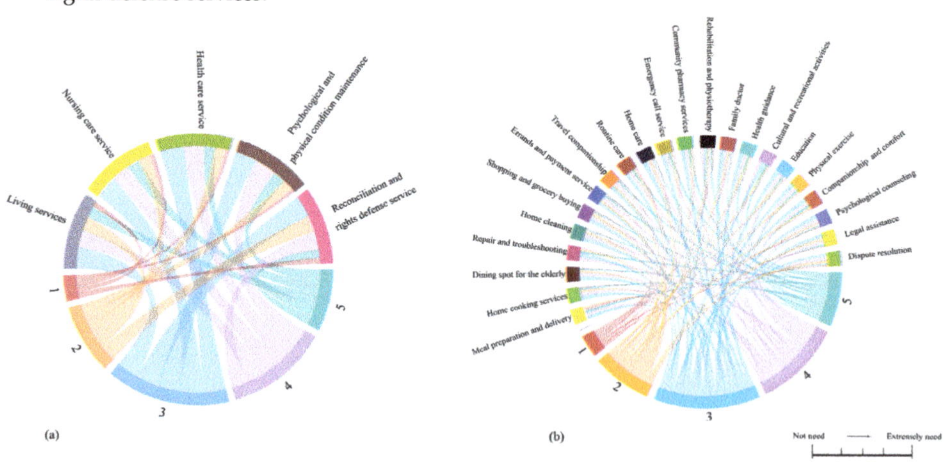

Figure 5. (**a**) Questionnaire responses for the 5 main categories; (**b**) the degree of demand for the 22 subcategories.

3.2. Factors That Affect the Demands for 22 Subcategories

Using the random forest regression model, the importance of all the indicators that influence the elderly demand for services were analyzed and the top 10 indicators were ranked (Figure 6). The highest R^2 in the model was 0.75 for travel companionship, followed by home cleaning at 0.64, rehabilitation and physiotherapy at 0.65, and psychological counseling at 0.6, all of which are above 0.6. Those with R^2 below 0.5 included dining for the elderly at 0.46, errands and payment service at 0.46 as well, cultural and recreational activities at 0.47, and dispute resolution at 0.49, indicating that the fit was relatively poor. Nevertheless, certain patterns were still discernible. In general, among all the influencing factors with the highest importance in the 22 subcategories of demand, 'mean distance from CECS to downtown landmark' occurred with the highest frequency, followed by "living arrangements" and "area ratio of all CECSs to subdistrict", belonging to characteristics of built environment and elderly themselves, with the former being more important. The second most important factors are "occupation before retirement", "living arrangements", "caregiver", "age", "gender", "monthly disposable income", "mean distance from CECS to downtown landmark", and "housing prices", which shows that the characteristics of the elderly themselves are more important than any others. The third most important factors were "educational background" and "mean distance from CECS to nearest subway stations". The first three positions of importance show that characteristics of the built environment, such as the accessibility of CECS, the locational advantage of the subdistricts where the elderly reside, and economic level affect their demand for elderly care services the most. Also, the elderly's attributes, such as age, living arrangements and economic status are also important factors.

Statistics on the frequency of the top three, top five, and top ten factors (Figure 7) show that among the top three, the most frequent were "mean distance from CECS to downtown landmark", "living arrangements", and "mean distance from CECS to nearest main roads", followed by "gender", "floor area ratio", "occupation before retirement", and "caregiver"; in the top five, the most important ones were the same as in the top 3 group, with the addition of "educational background"; and in the top 10, the most important one was still "mean distance from CECS to downtown landmark", followed by "living arrangements", "caregiver", "mean distance from CECS to nearest main roads", and "educational background". "Mean distance from CECS to downtown landmark" was the most important factor we found, which reflects the fact that the location of a CECS as well as its transportation accessibility are both keystones for the acquisition of elderly services in city centers.

Overall, the importance of activity characteristics among the attributes of the elderly was relatively low. Combining importance ranking with frequency of ranking, the relatively most important types of activities were "singing", "walking", and "Tai Chi", and the preferred place for activities was "outdoor space in the community". However, their importance was much lower than that of the elderly's own socio-economic attributes, and even less than built environment factors as well.

In the top three subcategories with the highest level of demand, cultural and recreational activities, repair and troubleshooting, and emergency call service, the most important influencing factors were, respectively, "housing prices", "mean distance from CECS to downtown landmark", and "occupation before retirement". In the subcategories with a higher level of demand, such as physical exercise and companionship and comfort, the most important factors were mainly the characteristics of CECS (6 in 7), except for "community pharmacy services", which corresponded to "living arrangements".

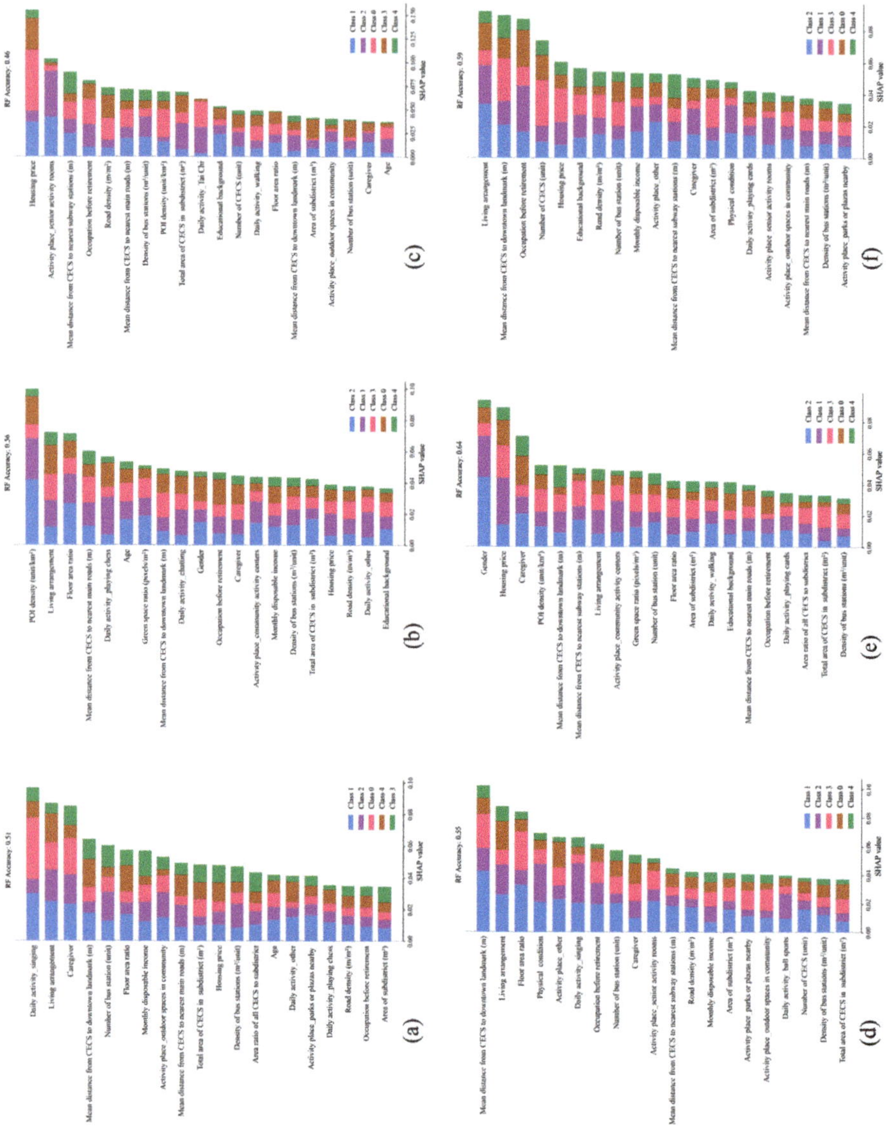

Figure 6. *Cont.*

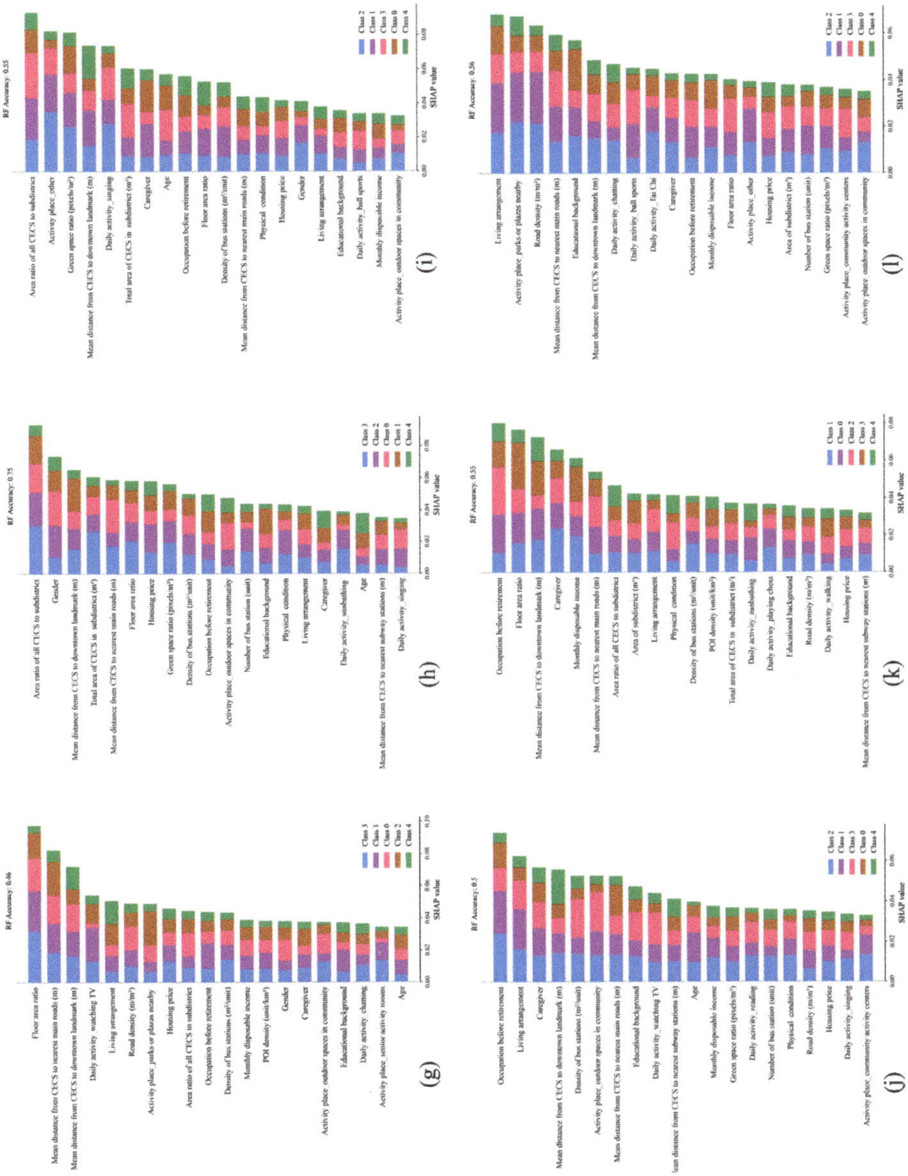

Figure 6. *Cont.*

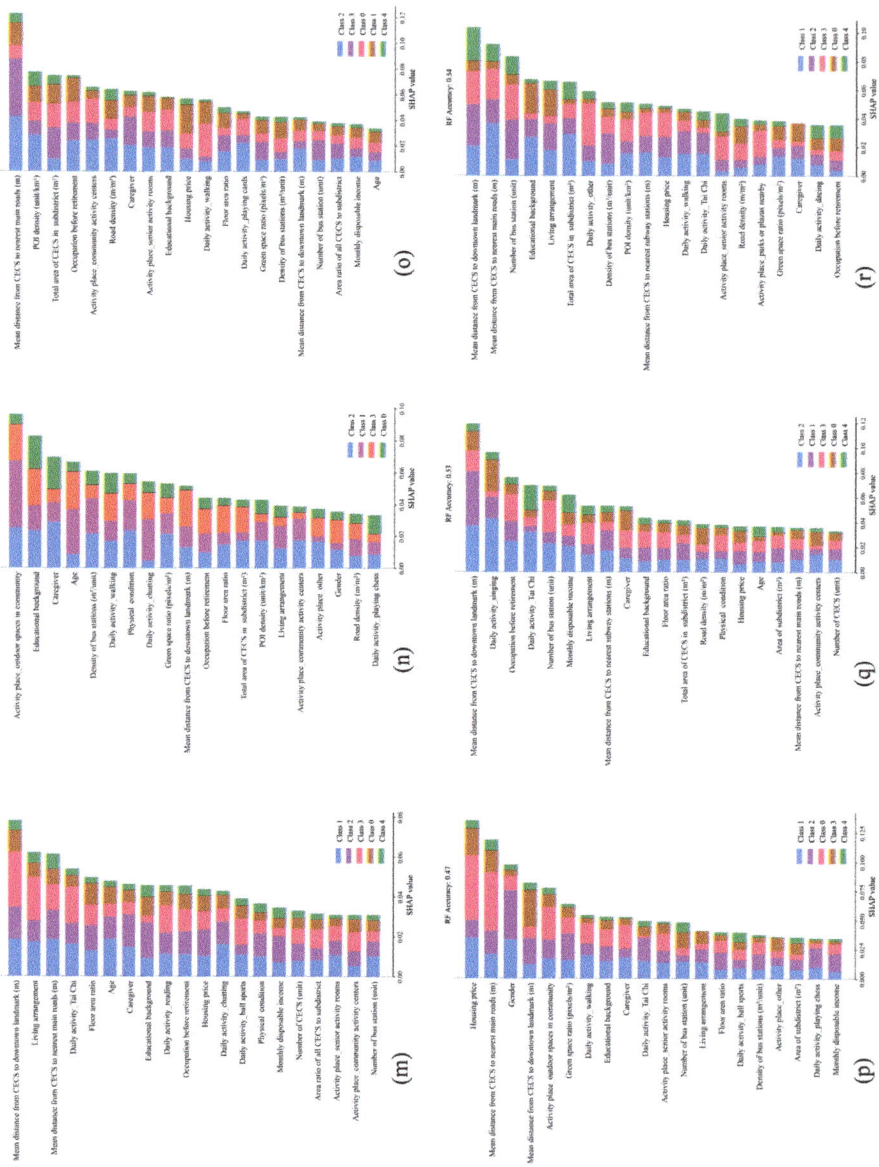

Figure 6. Cont.

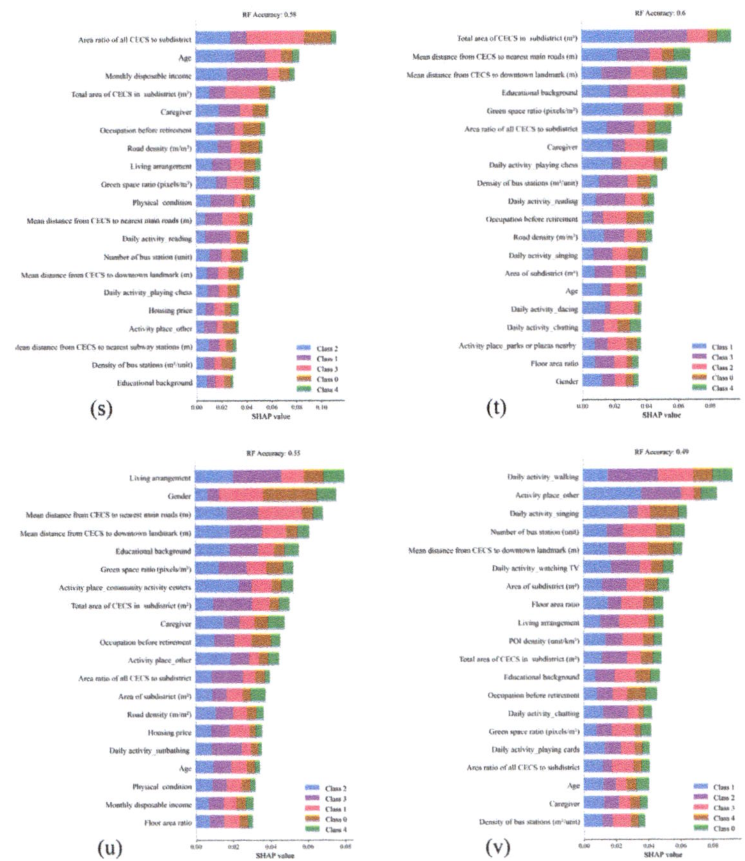

Figure 6. (**a**) Meal preparation and delivery; (**b**) home cooking services; (**c**) dining spot for the elderly; (**d**) repair and troubleshooting; (**e**) home cleaning; (**f**) shopping and grocery buying; (**g**) errands and payment service; (**h**) travel companionship; (**i**) routine care; (**j**) home care; (**k**) emergency call service; (**l**) community pharmacy services; (**m**) rehabilitation and physiotherapy; (**n**) family doctor; (**o**) health guidance; (**p**) cultural and recreational activities; (**q**) education; (**r**) physical exercise; (**s**) companionship and comfort; (**t**) psychological counseling; (**u**) legal assistance; (**v**) dispute resolution.

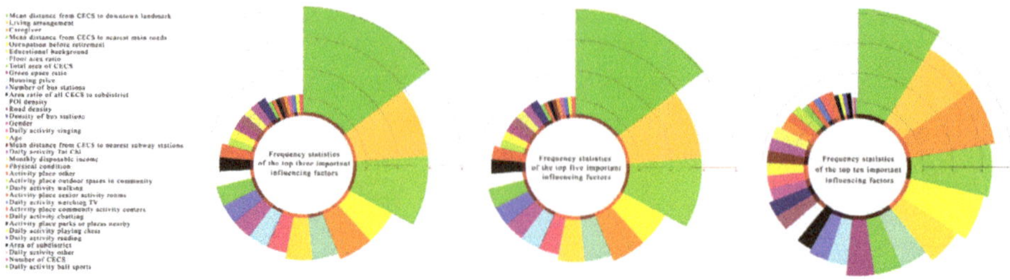

Figure 7. Frequency statistics of the top three, top five, and top ten most important influencing factors.

The top 10 important influencing factors in each of the 22 subcategories of demand all covered two types of characteristics, the built environment and the elderly themselves, but from the cross-sectional comparison of these subcategories, we find that with demand for travel companionship, eight factors stemmed from the built environment and only two factors from elderly people's own attributes, which may be due to the fact that the accessibility of the built environment is highly correlated with the ease of travel for older people. On the contrary, in the five subcategories of repair and troubleshooting, community pharmacy services, rehabilitation and physiotherapy, family doctor, and educational, factors from elderly's own attributes accounted for 7 out of 10. In addition, in 15 out of the 22 subcategories, the proportion of factors of the characteristics of CECS to the built environment was more than half, which shows again the importance of exclusive elderly resources such as in the built environment. Especially for psychological counseling, the proportions of characteristic of CECS in built environment features were 80%, and 75%, respectively, for both home care and legal assistance. These services are provided in CECSs, which suggests that elderly people who demand them should pay attention to the layout of CECSs in their areas.

3.3. Factors That Influence Five Categories of Demand

In the top three factors for the categories of living and health care service, the factors with the highest frequency were both characteristics of CECSs and living arrangements, but in nursing care service the highest frequency was for attributes of the elderly. In psychological and physical condition maintenance both highest-frequency factors were characteristics of the built environment. There was a single factor with highest frequency among the top five factors that was highly consistent in five categories, and it again was a built environment attribute. Finally, in the top 10, the factor with highest frequency in living service was a built environment attribute as well but in nursing care service was an elderly people attribute. The three remaining categories had the most important factors from both the built environment and the elderly people themselves (Table 6).

Demand for living services and reconciliation and rights defense service were highly consistent among the top 3, 5, and 10 most frequent factors, with mean distance from CECS to downtown landmark and living arrangements the most important. Psychological and physical condition maintenance, and health care services, were highly consistent as well, with the most important factors being accessibility of CECSs, followed by having a caregiver, and educational background. In this result, apart from the characteristics of CECS, no other built environment features appeared, nor did activity characteristics from older people's own attributes.

Table 6. Highest frequencies in the top three, five, and ten rankings of the factors in the five categories of demand (including ties).

Demand on Five Main Areas	Factors with the Highest Frequency in Top 3	Factors with the Highest Frequency in Top 5	Factors with the Highest Frequency in Top 10
Living services	Mean distance from CECS to downtown landmark/ Living arrangements	Mean distance from CECS to downtown landmark	Mean distance from CECS to downtown landmark
Nursing care service	Occupation before retirement	Mean distance from CECS to downtown landmark	Mean distance from CECS to downtown land-mark/Caregiver/Occupation before retirement
Health care service	Mean distance from CECS to nearest main roads/ Living arrangements	Mean distance from CECS to nearest main roads	Educational background/Caregiver
Psychological and physical condition maintenance	Mean distance from CECS to downtown landmark/Mean distance from CECS to nearest main roads	Mean distance from CECS to downtown landmark	Mean distance from CECS to downtown landmark/Educational background/Caregiver
Reconciliation and rights defense service	/	Mean distance from CECS to downtown landmark	Mean distance from CECS to downtown landmark/ Living arrangements

4. Discussion

4.1. Valuing and Using the Different Levels of Demand for Services

Descriptive statistics results for the 22 categories are consistent with some of the findings in previous studies, such as that healthy older people have a high demand for physical exercise and cultural and recreational activities [82–84], and disabled older people have a significant demand for routine care and emergency call service [85,86]. However, what was previously thought to be a high level of demand for living services among older people is contradicted by the results of the present study: with the exception of repair and troubleshooting, the level of demand for other types of living services was relatively low, especially for meal preparation and delivery, shopping and grocery buying, and travel companionship (mean < 3.20). This low level of demand for living services may be due to financial cost concerns [87], or it may be related to the fact that older people who are aging in place are predominantly active ones [41,42].

Situations in which older people need repair and troubleshooting are not common, but this work is both necessary and highly technical and difficult for older people to manage independently [88]. In addition, due to the current imperfect allocation of service resources and a lack of sufficient personnel, community pharmacy services in China cannot meet all the door-to-door medication needs of all elderly residents, and there is a severe shortage of both professional personnel and volunteers for companionship and comfort services [89,90].

In terms of the five main categories, older people's high demand for psychological and physical condition maintenance is consistent with previous research [91]. Studies have examined the varying frequency of use of social services by older people at risk of hospitalization who are aging in place and have found that social and cultural activities and catering services are frequently in demand [92]. Another study of the same type of older people found that personal care, social and cultural activities, and catering services were frequently in demand as well [93], demonstrating the importance of social and cultural activities. Therefore, CECFs should emphasize psychological and physical condition maintenance. Specifically, planning should be based on their subcategories, such as organizing health-related educational activities, festivals, and physical activities and providing psychological counseling or companionship in order to alleviate loneliness of the elderly. The demand for reconciliation and rights defense service is the lowest. Compared to other services, the frequency and quantity of demand for these issues was not high, and older people can obtain this sort of help through neighborhood committees, specialized law firms, or professional institutions [94,95]. Therefore, it is not necessary for CECF to allocate professional personnel and dedicated spaces for these services.

Nonetheless, a significant proportion of the demand for the 22 subcategories of services can be met through allocations in CECFs [96]. A first step should thus be to continue leveraging the service carrier function of CECFs, enhance their "in station" service capabilities and improve their door-to-door services. In particular, services with high demand levels identified in this study should be strengthened in terms of personnel and space to ensure adequate provision and avoid shortages [38]. For services with weak or no demand, however, further research should be conducted to understand the underlying reasons and appropriately reduce the supply for these services [97]. Furthermore, dynamic surveys should also be conducted to avoid fluctuations in future demand. In summary, elderly care services should be allocated based on elderly's demand for them [98].

4.2. Characteristics of the Built Environment Are Very Important in Influencing the Demands for Elderly Services

The important factors in the built environment dimension of affecting the demand for elderly services are the characteristics of the CECF, fully demonstrated that exclusive elderly care resources are very important in the built environment of "ageing in place", which has not been found in previous studies. Studies that focused on the role of CECFs specifically have found that due to poor planning in core areas of large cities these facilities cannot meet the demands for services [61,99]. For instance, irrational spatial distribution is the main reason for the imbalance in supply and demand for services in CECSs in Beijing, where it affects over 80% of CECSs [100]. The distribution of different types of CECFs is also unbalanced in the city, which has led to a deviation in the service capacity of facilities and resulted in service needs of the elderly remaining unmet [67]. Furthermore, there is a wide variation in the utilization of CECFs in most cities [101], as well as a general mismatch between supply and demand in terms of usage area [68]. These issues reflect a lack of adequate understanding of the demand of different types of older people during the layout process of CECFs.

The importance of accessibility is more prominent among the characteristic factors of CECFs, such as mean distance from CECS to downtown landmark and nearest roads in this study, which is consistent with the results of previous studies [61,102]. Moreover, the number of CECSs, the total area of CECSs, and area ratio of all CECSs to subdistrict, all reflect the coverage of CECSs, which itself has a certain impact on the fulfillment of demand for elderly services [100]. These results suggest that when planning CECFs, the accessibility of facilities should be considered first, followed by the setting of the number and size of CECFs in terms of subdistricts.

In addition to exclusive elderly care resources, the locational characteristics of the built environment, such as the level of development, accessibility, and amenity completeness, also have an important impact on the demand for elderly care services [103]. In this study, we found that floor area ratio, housing prices, and POI density are the most important in terms of the built environment. The former two reflect population density and urban location in a certain area [104]. For example, housing prices reflect the economic characteristics and the locational value of a subdistrict, as well residents' purchasing power and economic "capacity" [105], and POI density characterizes the level of public service facilities, which indirectly reflects the vitality of the urban built environment [106]. Other indicators of the built environment are also influential, such as green space ratio, which reflects the level of ecological friendliness within each subdistrict and is an important part of the livability of the built environment [107], and road density and density of bus stops both characterize the accessibility of the location [108]. The influence of these indicators suggests that although the built environment is difficult to change easily, it is still possible to optimize some built environment elements during the construction process of amenities for the elderly. Thus, favorable characteristics of the built environment in an area, such as good accessibility and economic development, should be considered when planning for the provision of elderly services [109].

4.3. Older People's Own Characteristics Have an Important Influence on Their Demand for Services

Living arrangements was the most important factor for three subcategories of demand (shopping and grocery buying, community pharmacy services, and legal assistance), and occupation before retirement was the most important for two (home care, emergency call service). Moreover, according to the frequency statistics in the top three, top five, and top ten importance ranking of influencing factors, it is clear that the frequency of the living arrangements was the highest, second only to the first characteristic element of CECSs. The next most important factors were caregiver, occupation before retirement, and educational background.

The important effect of living arrangements on the demand for elderly care services has already been studied before [48]. Firstly, the two types of elderly people, living alone and not living alone, have a significant effect on the demands on services [110]; and secondly, whether they live with children and grandchildren also affects their well-being. However, a very small number of elderly respondents living with nonrelatives nevertheless have high well-being and low levels of depression [111,112]. This evidence suggests that living arrangements determine who provides the care for elderly and consequently the extent to which the elderly require care from public resources [113], a result that is in line with those of this study.

The influence of having caregivers on the demand for elderly care service in this study was reflected in meal preparation and delivery, home cleaning, home care, and having a family doctor, which fully illustrates that the presence of a caregiver and the caregiver's attributes have a significant impact on older people's daily care, health maintenance, and nursing. Previous studies have found that the presence or absence of a caregiver can affect the services provided to older people as their health declines [114] and that differences in the type of caregiver, such as spouse, relative, or nursing staff, also affect quality of daily care and health care services [115,116].

Previous studies have also shown that occupation before retirement also has a significant impact on older people's choice of elderly care services, as it not only influences the level of pension but also affects older people's perceptions and acceptance of different types of services [117,118]. In addition, educational background also affects their demand for services primarily stemming from cognitive differences among the elderly that in turn lead to different lifestyles and affect the acceptance of certain services or the channels through which services are accessed [87].

The impact of activity characteristics on older people's demand for elderly care services has yet to be observed in previous studies. However, we discovered that among the attributes of the elderly, the importance of activity characteristics factors was relatively low, although specific activity types, such as singing, chatting, and Tai Chi, were found to exert significant influence on the demand for various elderly services. Those activities are reported to be the favorite and most frequently performed outdoor activities among urban Chinese older people [119,120]. Previous studies have also found that daily outdoor activities can promote the health and well-being of the elderly [121], and among activity places, outdoor spaces in the community are important for engaging in leisure activities, socializing, and exercising [122]. In general, active older people have a high demand for many different services, and this group also constitutes the main body of the aging-in-place population.

Therefore, given the important influence of the elderly's own attributes on their demand for elderly services, policy makers should focus on the demand of these different groups in order to better allocate service resources.

4.4. Limitations

There are several limitations to this study. First, due to the limited size of the research staff, only 242 valid questionnaires were collected, which is somewhat small. Second, this study did not stratify the elderly into different groups and set corresponding proportions of samples. Finally, although the data covered 24 subdistricts, full coverage of the core area of Beijing would have required 32. Therefore, follow-up studies need to expand the scope of sample collection, increase the sample size for the proportion of each type of elderly group, and maintain a balanced sample size in different subdistricts.

5. Conclusions

This study focused on the problem in which community and home-based elderly care services in the center area of China's large cities cannot meet the demand for elder services of elderly residents who are aging in place, especially in terms of insufficient variety of available services. Focusing on Beijing, the study explored the demand of elderly residents provided at the subdistrict level and analyzed the significance of multi-dimensional factors influencing them. It is found that elderly residents have the highest demand for psychological and physical condition maintenance, including cultural and recreational activities, physical exercise, and companionship and comfort, and the lowest demand for reconciliation and rights defense service, with legal assistance and dispute resolution. The demand for repair and troubleshooting in the living services category, as well as emergency call service in the nursing care services category, were also very strong. In regard to the influencing factors on the elderly choosing demand for elderly care services, built environment factors were found to be very important, especially the characteristics of CECFs, such as mean distance from CECSs to the downtown landmark and to main roads. The elderly's own attributes were also of some importance, daily life characteristics such as the living arrangements, having a caregiver, and occupation before retirement being the most important ones.

This study clarifies the differences in the demand for various types of elderly care services and identifies the significant factors influencing elderly's choice on service demands, thereby providing a reference for adjusting the types and quantities of elderly care services at the subdistrict level, and we hope it can help to promote the efficient allocation of community elderly care resources. In addition, the results highlight the importance of CECFs in meeting the demand of older people who are aging in place. Furthermore, the findings of this study may also provide a reference for future understanding of the needs of the elderly in urban core areas. By analyzing the basic characteristics of older residents within subdistricts, we may be able to deduce the intensity of their demand for different services, thereby allowing us to help improve service provision and facilities planning in order to create a more suitable environment for aging in place.

Author Contributions: Conceptualization, F.W.; methodology, F.W. and Y.Z. (Yuyang Zhang); software, Z.L. and Y.Z. (Yan Zhang); validation, Z.Z.; data acquisition, curation, and analysis, F.W., Z.L., Z.Z. and Y.Z. (Yan Zhang); visualization, Z.L.; writing—original draft, F.W. and Z.L.; writing—review and editing, B.Z. and Z.Z.; project administration, B.Z. and Y.Z. (Yuyang Zhang). All authors have read and agreed to the published version of the manuscript.

Funding: This research was funded by the Beijing Municipal Social Science Foundation, grant number 21SRC024.

Institutional Review Board Statement: The ethical approval of this study is not required according to the Chinese "Regulations on Ethical Review of Life Sciences and Medical Research Involving Humans".

Informed Consent Statement: Older adults who filled out the questionnaire were all informed and consented to the anonymous collection of their information.

Data Availability Statement: The raw data supporting the conclusions of this article will be made available by the authors upon request.

Conflicts of Interest: The authors declare no conflicts of interest.

References

1. Department of Economic Social Affairs, United Nations. World Population Prospects 2022: Summary of Results. Available online: https://www.un.org/development/desa/pd/sites/www.un.org.development.desa.pd/files/undesa_pd_2022_wpp_key-messages.pdf (accessed on 5 November 2024).
2. United Nations World Population Projected to Reach 9.8 Billion in 2050, and 11.2 Billion in 2100. Available online: https://www.un.org/en/desa/world-population-projected-reach-98-billion-2050-and-112-billion-2100#:~:text=Calendar-,World%20population%20projected%20to%20reach%209.8%20billion%20in%202050,%20and,Nations%20report%20being%20launched%20today (accessed on 6 May 2024).
3. United Nations World Population Prospects. 2022. Available online: https://population.un.org/wpp/ (accessed on 6 May 2024).
4. Population Division. Available online: https://www.un.org/development/desa/pd/ (accessed on 15 November 2024).
5. Smith, J.; Borchelt, M.; Maier, H.; Jopp, D. Health and Well–Being in the Young Old and Oldest Old. *J. Soc. Issues* **2002**, *58*, 715–732. [CrossRef]
6. Biedenharn, P.J.; Normoyle, J.B. Elderly Community Residents' Reactions to the Nursing Home: An Analysis of Nursing Home-Related Beliefs. *Gerontologist* **1991**, *31*, 107–115. [CrossRef]
7. Wiles, J.L.; Leibing, A.; Guberman, N.; Reeve, J.; Allen, R.E. The Meaning of "Aging in Place" to Older People. *Gerontologist* **2012**, *52*, 357–366. [CrossRef]
8. Harmand, M.G.-C.; Meillon, C.; Rullier, L.; Avila-Funes, J.-A.; Bergua, V.; Dartigues, J.-F.; Amieva, H. Cognitive Decline after Entering a Nursing Home: A 22-Year Follow-up Study of Institutionalized and Noninstitutionalized Elderly People. *J. Am. Med. Dir. Assoc.* **2014**, *15*, 504–508. [CrossRef] [PubMed]
9. Choi, Y.J. Understanding Aging in Place: Home and Community Features, Perceived Age-Friendliness of Community, and Intention toward Aging in Place. *Gerontologist* **2022**, *62*, 46–55. [CrossRef]
10. Lewis, C.; Buffel, T. Aging in Place and the Places of Aging: A Longitudinal Study. *J. Aging Stud.* **2020**, *54*, 100870. [CrossRef] [PubMed]
11. Tang, F.; Lee, Y. Social Support Networks and Expectations for Aging in Place and Moving. *Res. Aging* **2011**, *33*, 444–464. [CrossRef]
12. Wood, G.; Pykett, J.; Daw, P.; Agyapong-Badu, S.; Banchoff, A.; King, A.C.; Stathi, A. The Role of Urban Environments in Promoting Active and Healthy Aging: A Systematic Scoping Review of Citizen Science Approaches. *J. Urban Health* **2022**, *99*, 427–456. [CrossRef]
13. Jayantha, W.M.; Qian, Q.K.; Yi, C.O. Applicability of 'Aging in Place'in Redeveloped Public Rental Housing Estates in Hong Kong. *Cities* **2018**, *83*, 140–151. [CrossRef]
14. Vasunilashorn, S.; Steinman, B.A.; Liebig, P.S.; Pynoos, J. Others Aging in Place: Evolution of a Research Topic Whose Time Has Come. *J. Aging Res.* **2012**, *2012*, 120952. [CrossRef]
15. Yu, J.; Huang, W.; Kahana, E. Investigating Factors of Active Aging among Chinese Older Adults: A Machine Learning Approach. *Gerontologist* **2022**, *62*, 332–341. [CrossRef]

16. World Bank Open Data. Available online: https://data.worldbank.org (accessed on 15 November 2024).
17. Wang, P.P. The Decline in the Total Population Has Narrowed, and the Quality of the Population Has Continued to Improve. Available online: http://www.ce.cn/xwzx/gnsz/gdxw/202501/17/t20250117_39269947.shtml (accessed on 17 January 2025).
18. Ai, J.; Feng, J.; Yu, Y. Elderly Care Provision and the Impact on Caregiver Health in China. *China World Econ.* **2022**, *30*, 206–226. [CrossRef]
19. Chen, L.; Han, W.-J. Shanghai: Front-Runner of Community-Based Eldercare in China. *J. Aging Soc. Policy* **2016**, *28*, 292–307. [CrossRef] [PubMed]
20. Zhou, J.; Walker, A. The Impact of Community Care Services on the Preference for Ageing in Place in Urban China. *Health Soc. Care Community* **2021**, *29*, 1041–1050. [CrossRef]
21. Henderson, E.J.; Caplan, G.A. Home Sweet Home? Community Care for Older People in Australia. *J. Am. Med. Dir. Assoc.* **2008**, *9*, 88–94. [CrossRef]
22. Dong, Y. Spatial Countermeasures for Ageing-in-Place in Big Cities. *Urban Plan. Forum* **2015**, *4*, 83–90.
23. He, Y.; Wei, B.; Li, Y. The Impact of Using Community Home-Based Elderly Care Services on Older Adults' Self-Reported Health: Fresh Evidence from China. *Front. Public Health* **2023**, *11*, 1257463. [CrossRef]
24. Zhang, Z.; Mao, Y.; Shui, Y.; Deng, R.; Hu, Y. Do Community Home-Based Elderly Care Services Improve Life Satisfaction of Chinese Older Adults? An Empirical Analysis Based on the 2018 CLHLS Dataset. *Int. J. Environ. Res. Public Health* **2022**, *19*, 15462. [CrossRef]
25. Misu, Y.; Hayashi, S.; Iwai, N.; Yamamoto, T. Factors Affecting the Life Satisfaction of Older People with Care Needs Who Live at Home. *Geriatrics* **2022**, *7*, 117. [CrossRef]
26. Gu, S.; Jia, C.; Shen, F.; Wang, X.; Wang, X.; Gu, H. The Effects of Community-Based Home Health Care on the Physical and Mental Health of Older Adults with Chronic Diseases. *Qual. Life Res.* **2024**, *33*, 691–703. [CrossRef]
27. Hu, J.; Zhang, Y.; Wang, L.; Shi, V. An Evaluation Index System of Basic Elderly Care Services Based on the Perspective of Accessibility. *Int. J. Environ. Res. Public Health* **2022**, *19*, 4256. [CrossRef] [PubMed]
28. Xia, C. Community-Based Elderly Care Services in China: An Analysis Based on the 2018 Wave of the CLHLS Survey. *China Popul. Dev. Stud.* **2020**, *3*, 352–367. [CrossRef]
29. Zhao, R.; Chen, R.; Ma, Y. The Problems and Strategies of Urban Community Service System for the Elderly. *Chin. J. Gerontol.* **2015**, *35*, 3399–3401.
30. Gong, N.; Meng, Y.; Hu, Q.; Du, Q.; Wu, X.; Zou, W.; Zhu, M.; Chen, J.; Luo, L.; Cheng, Y.; et al. Obstacles to Access to Community Care in Urban Senior-Only Households: A Qualitative Study. *BMC Geriatr.* **2022**, *22*, 122. [CrossRef] [PubMed]
31. Qin, S.; Cheng, Y.; Zhang, H.; Ding, Y. Home/Community-Based Medical and Elderly Care Services Utilization in China: A Cross-Sectional Study from the Middle-Aged and Elderly Population. *Healthcare* **2023**, *11*, 2431. [CrossRef]
32. Wang, K.; Ke, Y.; Sankaran, S.; Xia, B. Problems in the Home and Community-Based Long-Term Care for the Elderly in China: A Content Analysis of News Coverage. *Int. J. Health Plann. Manag.* **2021**, *36*, 1727–1741. [CrossRef]
33. Liu, Y.; Zhang, X.; Zhang, Y. Service Supply Chain Optimization of Community Elderly Care-A Case Study in Beijing. In *Proceedings of the International Conference on Logistics, Informatics and Service Sciences, Beijing, China, 21–23 July*; Springer: Berlin/Heidelberg, Germany, 2022; pp. 487–511.
34. Zheng, J.; Xu, J.; Zhuo, L. Impacts of Health Condition on Demand of Home-Based Pension Service among Community Elderly. *Chin. J. Public Health* **2020**, *36*, 545–548.
35. Gao, X.; Yan, B.; Ji, J. Urban Elders' Desirable Caring Patterns and Its Rationality: A Decision Tree Analysis. *Prog. Geogr.* **2012**, *31*, 1274–1281.
36. Wu, D.; Gao, X.; Xie, Z.; Xu, Z. Understanding the Unmet Needs among Community-Dwelling Disabled Older People from a Linkage Perspective. *Int. J. Environ. Res. Public Health* **2021**, *18*, 389. [CrossRef]
37. Xiang, L.; Yu, A.T.; Tan, Y.; Shan, X.; Shen, Q. Senior Citizens' Requirements of Services Provided by Community-Based Care Facilities: A China Study. *Facilities* **2020**, *38*, 52–71. [CrossRef]
38. Longduoqi, A.; Ma, H.; Wang, M.; Yang, B. Research on Urban Community Elderly Care Facility Based on Quality of Life by SEM: Cases Study of Three Types of Communities in Shenzhen, China. *Sustainability* **2022**, *14*, 9661. [CrossRef]
39. de Stampa, M.; Bagaragaza, E.; Cerase, V.; Lys, E.; Alitta, Q.; Aegerter, P.; Henrard, J.C.; Ankri, J. Predictive Factors Leading to Not Living at Home in a Very Frail Community Dwelling Older Population. *Int. J. Gerontol.* **2019**, *13*, 23–27.
40. Cao, M.; Guo, X.; Yu, H.; Chen, L.; McDonald, T.A. Chinese Community-Dwelling Elders' Needs: Promoting Ageing in Place. *Int. Nurs. Rev.* **2014**, *61*, 327–335. [CrossRef] [PubMed]
41. Hao, X. Health Status and Demands of the Elderly Population for Community Health Services in Beijing Communities. *Chin. Gen. Pract.* **2010**, *13*, 2850–2852.

42. Xu, W. Health Status of Elders in Urban Communities and Their Demands for Health Services in China. *Chin. Gen. Pract.* **2010**, *13*, 2846–2849.
43. Feng, Z.; Glinskaya, E.; Chen, H.; Gong, S.; Qiu, Y.; Xu, J.; Yip, W. Long-Term Care System for Older Adults in China: Policy Landscape, Challenges, and Future Prospects. *Lancet* **2020**, *396*, 1362–1372. [CrossRef]
44. Zhang, Y.; Yeager, V.A.; Hou, S. The Impact of Community-Based Supports and Services on Quality of Life among the Elderly in China: A Longitudinal Study. *J. Appl. Gerontol.* **2018**, *37*, 1244–1269. [CrossRef]
45. Huang, G.; Guo, F.; Chen, G. Utilization of Home-/Community-Based Care Services: The Current Experience and the Intention for Future Utilization in Urban China. *Popul. Res. Policy Rev.* **2023**, *42*, 61. [CrossRef]
46. Xin, S.; Yali, Z.; Zihua, X.; Mei, Z.; Jing, J.; Xiaojun, L.; Dongyan, J. Needs of Community-Based Home Care Services for Mongolian Elderly People in Hohhot. *J. Nurs. Sci.* **2020**, *35*, 86–89.
47. House, J.S.; Lepkowski, J.M.; Kinney, A.M.; Mero, R.P.; Kessler, R.C.; Herzog, A.R. The Social Stratification of Aging and Health. *J. Health Soc. Behav.* **1994**, *35*, 213–234. [CrossRef]
48. Hughes, M.E.; Waite, L.J. Health in Household Context: Living Arrangements and Health in Late Middle Age. *J. Health Soc. Behav.* **2002**, *43*, 1.
49. Stringhini, S.; Sabia, S.; Shipley, M.; Brunner, E.; Nabi, H.; Kivimaki, M.; Singh-Manoux, A. Association of Socioeconomic Position with Health Behaviors and Mortality. *JAMA* **2010**, *303*, 1159–1166. [PubMed]
50. Clarke, P.; Erreygers, G. Defining and Measuring Health Poverty. *Soc. Sci. Med.* **2020**, *244*, 112633. [PubMed]
51. Lam, C.L.K.; Guo, V.Y.; Wong, C.K.H.; Yu, E.Y.T.; Fung, C.S.C. Poverty and Health-Related Quality of Life of People Living in Hong Kong: Comparison of Individuals from Low-Income Families and the General Population. *J. Public Health* **2017**, *39*, 258–265.
52. Zhan, H.J.; Liu, G.; Guan, X.; Bai, H. Recent Developments in Institutional Elder Care in China: Changing Concepts and Attitudes. *J. Aging Soc. Policy* **2006**, *18*, 85–108.
53. Wang, C.; Rao, K.; Wu, S.; Liu, Q. Health Care in China: Improvement, Challenges, and Reform. *Chest* **2013**, *143*, 524–531.
54. Almeida, A.P.S.C.; Nunes, B.P.; Duro, S.M.S.; Facchini, L.A. Socioeconomic Determinants of Access to Health Services among Older Adults: A Systematic Review. *Rev. Saude Publica* **2017**, *51*, 50.
55. Zeng, Y.; Que, S.; Lin, C.; Fang, Y. The Expected Demand for Elderly Care Services and Anticipated Living Arrangements among the Oldest Old in China Based on the Andersen Model. *Front. Public Health* **2021**, *9*, 715586.
56. Kong, F.; Xu, L.; Kong, M.; Li, S.; Zhou, C.; Zhang, J.; Ai, B. Association between Socioeconomic Status, Physical Health and Need for Long-Term Care among the Chinese Elderly. *Int. J. Environ. Res. Public Health* **2019**, *16*, 2124. [CrossRef]
57. Zhang, J.; Xu, X.; Yang, L.; Wang, J. Met and Unmet Care Needs of Home-Living People with Dementia in China: An Observational Study Using the Camberwell Assessment of Need for the Elderly. *Geriatr. Gerontol. Int.* **2021**, *21*, 102–107.
58. Zeng, Y.; Lin, C.; Fang, Y. Factors Influencing the Demands for Home and Community-Based Care Services among Older Adults Living Alone in China. *J. Aging Soc. Policy* **2023**, *36*, 1045–1061. [CrossRef] [PubMed]
59. Menec, V.H. The Relation between Everyday Activities and Successful Aging: A 6-Year Longitudinal Study. *J. Gerontol. B Psychol. Sci. Soc. Sci.* **2003**, *58*, S74–S82. [PubMed]
60. Zhao, X.; Yu, J.; Liu, N. Relationship between Specific Leisure Activities and Successful Aging among Older Adults. *J. Exerc. Sci. Fit.* **2023**, *21*, 111–118. [PubMed]
61. Ning, W.; Yang, Y.; Lu, M.; Han, X. Equity in Walking Access to Community Home Care Facility Resources for Elderly with Different Mobility: A Case Study of Lianhu District, Xi'an. *PLoS ONE* **2022**, *17*, e0277310.
62. Chen, S.; Wang, T.; Bao, Z.; Lou, V. A Path Analysis of the Effect of Neighborhood Built Environment on Public Health of Older Adults: A Hong Kong Study. *Front. Public Health* **2022**, *10*, 861836.
63. Feng, J. The Influence of Built Environment on Travel Behavior of the Elderly in Urban China. *Transp. Res. Part D Transp. Environ.* **2017**, *52*, 619–633.
64. Zhang, F.; Loo, B.P.; Wang, B. Aging in Place: From the Neighborhood Environment, Sense of Community, to Life Satisfaction. *Ann. Am. Assoc. Geogr.* **2022**, *112*, 1484–1499.
65. Mu, J.; Kang, J. Indoor Environmental Quality of Residential Elderly Care Facilities in Northeast China. *Front. Public Health* **2022**, *10*, 860976.
66. Chang, F.; Wang, L.; Ma, Y.; Yan, C.; Liu, H. Do Urban Public Service Facilities Match Population Demand? Assessment Based on Community Life Circle. *Prog. Geogr.* **2021**, *40*, 607–619. [CrossRef]
67. Cheng, T.; Liu, C.; Yang, H.; Wang, N.; Liu, Y. From Service Capacity to Spatial Equity: Exploring a Multi-Stage Decision-Making Approach for Optimizing Elderly-Care Facility Distribution in the City Centre of Tianjin, China. *Sustain. Cities Soc.* **2022**, *85*, 104076.
68. Wu, S.; Xia, X.; Wang, X.; Zhou, J.; Zhan, W.; Wu, W. Functional Space Optimization of Senior Service Centers in Urban Old Communities: An Example from Ningbo, China. *Fresen Environ. Bull.* **2022**, *31*, 9390–9402.

69. Zhang, J.; Liu, H.; Qi, Y.; Li, X. Living Environment and Life Satisfaction of Aged Population in Beijing Municipality. *Prog. Geogr.* **2015**, *34*, 1628–1636.
70. Beijing Municipal Office of the Working Committee for the Elderly. *Beijing Association of the Elderly*; China Institute for Public Welfare of Beijing Normal University: Beijing, China, 2021.
71. Wang, Y.; Wang, X.; Liu, Y.; Zhang, H.; Zhao, F. The Current Status of Home-Based Pension Service Demand and Supply in Changchun Community. *Chin. Nurs. Manag.* **2016**, *16*, 389–391.
72. Meng, D.; Zhang, M.; Xu, G.; Lin, D.; Ding, Y. Quality Attributes Analysis of Home Care Requirements of Urban Elderly Using Kano's Model. *Nurs. J. Chin. Peoples Lib. Army* **2020**, *37*, 35–38.
73. Chen, L.; Wen, W. Complement or Substitution: Family Care and Community Home-Based Care Services. *Chin. J. Health Policy* **2021**, *14*, 35–42.
74. Gan, L.; Zhou, L.; Xu, X. Research on the Construction of Evaluation Index System for High-Quality Development of Community Home-Based Elderly Care Services. *Chin. J. Health Policy* **2023**, *16*, 40–47.
75. Grimby, G.; Grimby, A.; Frändin, K.; Wiklund, I. Physically Fit and Active Elderly People Have a Higher Quality of Life. *Scand. J. Med. Sci. Sports* **1992**, *2*, 225–230.
76. Zhao, S.; Miao, M.; Wang, Q.; Yang, H.; Zhao, H.; Yao, H.; Liu, F.; Wang, X. Demand and Supply of Community-Based Care Services for the Elderly in China and Its Influencing Factors. *Chin. J. Geriatr.* **2023**, *12*, 92–97.
77. Huang, X.; Gong, P.; White, M.; Zhang, B. Research on Spatial Distribution Characteristics and Influencing Factors of Pension Resources in Shanghai Community-Life Circle. *ISPRS Int. J. Geo-Inf.* **2022**, *11*, 518. [CrossRef]
78. Ni'na, L.; Xixi, Z. Home- and Community-Based Care System in Urban China: Taking Beijing as an Example. *Sci. Res. Aging* **2021**, *9*, 29–41.
79. Zhang, Y.; Yu, D.; Zhao, H.; Zhang, B.; Li, Y.; Zhang, J. Chasing the Heat: Unraveling Urban Hyperlocal Air Temperature Mapping with Mobile Sensing and Machine Learning. *Sci. Total Environ.* **2024**, *927*, 172168. [CrossRef]
80. Zhang, Y.; Zhao, H.; Li, Y.; Long, Y.; Liang, W. Predicting Highly Dynamic Traffic Noise Using Rotating Mobile Monitoring and Machine Learning Method. *Environ. Res.* **2023**, *229*, 115896. [CrossRef] [PubMed]
81. Zhang, Y.; Ma, W.; Du, P.; Li, S.; Gao, K.; Wang, Y.; Liu, Y.; Zhang, B.; Yu, D.; Zhang, J.; et al. Powering the Future: Unraveling Residential Building Characteristics for Accurate Prediction of Total Electricity Consumption during Summer Heat. *Appl. Energy* **2024**, *376*, 124146. [CrossRef]
82. Zhang, W.; Feng, Q.; Fong, J.H.; Chen, H. Leisure Participation and Cognitive Impairment among Healthy Older Adults in China. *Res. Aging* **2023**, *45*, 185–197. [CrossRef] [PubMed]
83. Cheung, M.; Ting, W.; Chan, L.; Ho, K.; Chan, W. Leisure Participation and Health-Related Quality of Life of Community Dwelling Elders in Hong Kong. *Asian J. Gerontol. Geriatr.* **2009**, *4*, 15–23.
84. Wei, L.; Hu, Y.; Tao, Y.; Hu, R.; Zhang, L. The Effects of Physical Exercise on the Quality of Life of Healthy Older Adults in China: A Systematic Review. *Front. Psychol.* **2022**, *13*, 895373. [CrossRef]
85. Jiang, H.; Xiao, S.; Hu, H.; He, H. Study on the Measurement and Influencing Factors of Care Service Demand of Disabled Elderly in Urban and Rural China. *Int. J. Environ. Res. Public Health* **2022**, *19*, 11112. [CrossRef]
86. Wu, F.; Fan, H.; Xiao, S.; Yang, X.; Xiao, Y.; Wu, Y. The Demand for Care and Its Influencing Factors in Community-Dwelling Disabled Elderly in Beijing. *Chin. J. Nurs.* **2018**, *53*, 841–845.
87. Marinelli, M.; Zhang, J.; Ying, Z. Present and Future Trends of Sustainable Eldercare Services in China. *J. Popul. Ageing* **2023**, *16*, 589–617. [CrossRef]
88. Coleman, T.; Kearns, R.A.; Wiles, J. Older Adults' Experiences of Home Maintenance Issues and Opportunities to Maintain Ageing in Place. *Hous. Stud.* **2016**, *31*, 964–983. [CrossRef]
89. Tang, L. The Chinese Community Patient's Life Satisfaction, Assessment of Community Medical Service, and Trust in Community Health Delivery System. *Health Qual. Life Outcomes* **2013**, *11*, 1–12. [CrossRef] [PubMed]
90. Huang, L.; Zhang, F.; Guo, L.; Chen, Y.; Feng, M.; You, Y.; Zhang, L.; Jiang, Z.; Liu, Y. Experiences and Expectations of Receiving Volunteer Services among Home-Based Elderly in Chinese Urban Areas: A Qualitative Study. *Health Expect.* **2022**, *25*, 3164–3174. [CrossRef] [PubMed]
91. Fang, Y.; Wang, H.; Ouyang, Z. An Overview Report on the Situation of Elderly People in Urban and Rural China: 2000–2015. *China Popul. Dev. Stud.* **2019**, *2*, 323–345. [CrossRef]
92. van Bilsen, P.M.; Hamers, J.P.; Groot, W.; Spreeuwenberg, C. The Use of Community-Based Social Services by Elderly People at Risk of Institutionalization: An Evaluation. *Health Policy* **2008**, *87*, 285–295. [CrossRef] [PubMed]
93. van Bilsen, P.; Hamers, J.; Don, A.; Groot, W.; Spreeuwenberg, C. The Use of Social Services by Community-Dwelling Older Persons Who Are at Risk of Institutionalization: A Survey. *Eur. J. Ageing* **2010**, *7*, 101–109. [CrossRef]

94. Tang, B. Deliberating Governance in Chinese Urban Communities. *China J.* **2015**, *73*, 84–107. [CrossRef]
95. Hu, J. *Disputes Resolution in Urban Communities in Contemporary China*; Springer: Berlin/Heidelberg, Germany, 2020.
96. Wen, F.; Zhang, Y.; Du, P.; Zhang, Z.; Zhang, B.; Zhang, Y. Factors Influencing the Usage Frequency of Community Elderly Care Facilities and Their Functional Spaces: A Multilevel Based Study. *Buildings* **2024**, *14*, 1827. [CrossRef]
97. Shi, J.; Hua, W.; Tang, D.; Xu, K.; Xu, Q. A Study on Supply-Demand Satisfaction of Community-Based Senior Care Combined with the Psychological Perception of the Elderly. *Healthcare* **2021**, *9*, 643. [CrossRef]
98. Lixiong, Y. Community-Based Elderly Care in Beijing: Status and Prospects. In *Analysis of the Development of Beijing, 2019*; Springer: Berlin/Heidelberg, Germany, 2021; pp. 291–325.
99. Yang, Z.; Jiang, Y.; Wang, M.; Zeng, H. Current Status and Challenges of Community-Based Elderly Care Centers in Chongqing, China: A Cross-Sectional Study. *Risk Manag. Healthc. Policy* **2020**, *13*, 2975–2983. [CrossRef]
100. Li, H.; Liang, Z. Research on Spatial Equilibrium of Urban Community Elderly Care Facilities and Its Configuration. *J. Geo-Inf. Sci.* **2021**, *23*, 467–478.
101. Zhang, H.; Zhang, S. Reasons for Underutilization of Community Care Facilities for the Elderly in China. *BMC Geriatr.* **2024**, *24*, 791.
102. Zhou, S.; Cheng, Y.; Xiao, M.; Bao, X. Assessing the Location of Public-and-Community Facilities for the Elderly in Beijing, China. *GeoJournal* **2013**, *78*, 539–551.
103. Wang, N.; Ma, W. Evaluation Study on the Layout of Community Senior Living Facilities in the Main Area of Shanghai from the Perspective of Senior Life Circle. *Sustainability* **2023**, *15*, 10362. [CrossRef]
104. Yuan, Y.; Lei, M.; Hongqing, L.; Xiaofeng, Z. Constructing Index for the Assessment of Urban Residential Land Efficiency Using Location-Based Big Data. *J. Geo-Inf. Sci.* **2022**, *24*, 235–248.
105. Gong, Y.; Boelhouwer, P.; de Haan, J. Interurban House Price Gradient: Effect of Urban Hierarchy Distance on House Prices. *Urban Stud.* **2016**, *53*, 3317–3335. [CrossRef]
106. Jin, A.; Ge, Y.; Zhang, S. Spatial Characteristics of Multidimensional Urban Vitality and Its Impact Mechanisms by the Built Environment. *Land* **2024**, *13*, 991. [CrossRef]
107. Pan, Y.; Wang, Y.; Wang, Y.; Xie, Y.; Dong, J.; Liu, M. Spatiotemporal Dynamics of the Suitability for Ecological Livability of Green Spaces in the Central Yunnan Urban Agglomeration. *Sustainability* **2023**, *15*, 15964. [CrossRef]
108. Ma, J.; Huang, H.; Liu, D. Influences of Spatial Accessibility and Service Capacity on the Utilization of Elderly-Care Facilities: A Case Study of the Main Urban Area of Chongqing. *Int. J. Environ. Res. Public Health* **2023**, *20*, 4730. [CrossRef] [PubMed]
109. Zhao, P.; Zhao, Z.; Liao, C.; Fang, Y.; Liu, Y. Urban Park Accessibility and the Mental Health of Older Adults: A Case Study of Haidian District, Beijing. *Leis. Stud.* **2023**, *42*, 235–252. [CrossRef]
110. Chou, K.-L.; Chi, I. Comparison between Elderly Chinese Living Alone and Those Living with Others. *J. Gerontol. Soc. Work* **2000**, *33*, 51–66. [CrossRef]
111. Wang, J.; Chen, T.; Han, B. Does Co-Residence with Adult Children Associate with Better Psychological Well-Being among the Oldest Old in China? *Aging Ment. Health* **2014**, *18*, 232–239. [CrossRef] [PubMed]
112. Ren, Q.; Treiman, D.J. Living Arrangements of the Elderly in China and Consequences for Their Emotional Well-Being. *Chin. Sociol. Rev.* **2015**, *47*, 255–286. [CrossRef]
113. Meng, D.; Xu, G.; He, L.; Zhang, M.; Lin, D. What Determines the Preference for Future Living Arrangements of Middle-Aged and Older People in Urban China? *PLoS ONE* **2017**, *12*, e0180764. [CrossRef]
114. Soysal, P.; Rossi, F.; Portera, D.; Smith, L.; Yang, L.; Isik, A.T. The Role of Caregivers in the Care of Older People. In *The Role of Family Physicians in Older People Care*; Springer: Berlin, Germany, 2022; pp. 275–288.
115. Denton, M. The Linkages between Informal and Formal Care of the Elderly. *Can. J. Aging Rev. Can. Vieil.* **1997**, *16*, 30–50. [CrossRef]
116. Chida, I.; Pawar, M.; Mungai, N.W. Rethinking the Conceptualization of Formal and Informal Care for the Elderly. In *The Palgrave Handbook of Global Social Problems*; Springer: Berlin/Heidelberg, Germany, 2023; pp. 1–22.
117. Wu, L. Inequality of Pension Arrangements among Different Segments of the Labor Force in China. *J. Aging Soc. Policy* **2013**, *25*, 181–196. [CrossRef]
118. Wang, M.; Yang, Y.; Jin, S.; Gu, L.; Zhang, H. Social and Cultural Factors That Influence Residential Location Choice of Urban Senior Citizens in China–The Case of Chengdu City. *Habitat. Int.* **2016**, *53*, 55–65. [CrossRef]
119. Cai, X.; Huang, Y.; Zhang, B. Singing Together in the Park: Older Peoples' Wellbeing and the Singingscape in Guangzhou, China. *Emot. Space Soc.* **2023**, *47*, 100947. [CrossRef]
120. Zhang, B.; Huang, Y. Practicing Taichi Together in the Park: A Case Study of Taichiscape and Older People's Wellbeing. *Popul. Space Place* **2023**, *29*, e2696. [CrossRef]

121. Kerr, J.; Marshall, S.; Godbole, S.; Neukam, S.; Crist, K.; Wasilenko, K.; Golshan, S.; Buchner, D. The Relationship between Outdoor Activity and Health in Older Adults Using GPS. *Int. J. Environ. Res. Public Health* **2012**, *9*, 4615–4625. [CrossRef]
122. Lu, C.; Wu, W.; Han, D. Understanding the Spatial Distribution and Behavior of Elderly Residents in Age-Friendly Communities: An Analysis of Outdoor Space Features in Hangzhou, China. *Sustainability* **2023**, *15*, 10703. [CrossRef]

applied sciences

MDPI

Article

Experimental Evaluation of Precise Placement with Pushing Primitive Based on Cartesian Force Control

Jinseong Park *, Jeong-Jung Kim and Doo-Yeol Koh

Department of Artificial Intelligence Machinery, Korea Institute of Machinery & Materials, Daejeon 34103, Republic of Korea; rightcore@kimm.re.kr (J.-J.K.); dyk@kimm.re.kr (D.-Y.K.)
* Correspondence: jspark2090@kimm.re.kr; Tel.: +82-42-868-7181

Abstract: In-hand manipulation with Cartesian-force-control-based pushing primitives is introduced to achieve the precise placement of an object in a desired position at a manufacturing site. In the bin picking process, achieving the desired grasping posture is challenging due to limitations in the sensing and control of the robotic arm, interference from clustered objects, and unintended collisions, which hinder achieving the planned pose. Even under such conditions, in cases that require precise operations, such as manufacturing processes, maintaining a desired placement posture is crucial for the precise placement of objects into the machine slot. In this paper, a pushing primitive incorporating force feedback control is applied to ensure that the gripper is consistently positioned at the edge of the grasped object regardless of the initial grasping position by utilizing the surrounding environment of the processing machine. Modeling the exact contact friction between the gripper and the grasped object is challenging; therefore, instead of relying on a motion planning approach, we addressed the problem using a control method that leverages feedback from the external force information of the robot manipulator. Additional sensors such as external cameras or tactile sensors in the gripper are not required. The pushing primitive is executed by applying a force greater than the frictional force between the gripper and the grasped object, leveraging the surrounding environment. Experimental verification confirmed that the proposed method achieves precise placement into the machine slot, regardless of initial grasping positions. It also proved to be effective on an actual testbed.

Keywords: Cartesian force control; pushing primitives; in-hand manipulation; precise placement

1. Introduction

Pick-and-place operations constitute one of the main tasks of the robotic manipulation used to transfer objects from one process to another to realize factory automation. The successful grasping of objects placed in an unstructured manner in a bin or on a shelf and the placement of them in a desired posture is essential to expand the practical applications of robotic manipulation [1]. In various industrial processes, including logistics processes, the main focus has been on picking objects accurately [2–5]. However, in operations such as inserting parts into a machine slot within manufacturing workflows, precise placement is often even more critical. For secure picking, accurately recognizing the current pose of the object and determining an appropriate grasp position to hold it securely are key challenges. Consequently, research has focused on AI-based perception and decision-making problems, including clearing surrounding objects and regrasping [6–8]. On the other hand, accurate placing can be complicated by various issues, such as the imprecise recognition of the object's grasping position, caused by limitations in recognition algorithms or adverse imaging conditions, and difficulties in achieving the desired grasping posture due to limitations in the control accuracy of the robot manipulator. Therefore, it is necessary to address the challenges of accurate placement not only through AI-based recognition approaches but also by utilizing motion planning [9–14] or control techniques [15–21].

One of the simplest approaches used to achieve precise placement in robotic manipulation in manufacturing sites is to adopt manipulation assist systems, such as customized grippers, jigs, or part feeders, to make the environment more structured [1]. However, the use of such specialized systems requires additional space and cost, which can be a burden for industries. Regrasping, where the object is placed on a surface and grasped again with the desired posture, is another possible approach [22,23]. However, it is not always feasible to place the object onto a surface for regrasping. On the other hand, in-hand manipulation with dexterity can be exploited. This method can be categorized based on whether it utilizes resources intrinsic to the hand (intrinsic dexterity) or extrinsic to the hand (extrinsic dexterity). From the perspective of intrinsic dexterity, in-hand manipulation with finely position- and force-controlled articulated hands has been developed to achieve the dexterity and flexibility of human hands [24]. A multifingered hand with a passivity-based object-level controller using a virtual object model has also been introduced [25]. Additionally, a comprehensive survey of theoretical research on grasping and manipulation, covering topics such as grasp position optimization and impedance control, was provided [26]. Extrinsic dexterity with a simple gripper, on the other hand, utilizes external sources such as pushing with external contacts, gravity, or dynamic motions [9,10]. Prehensile pushing against the environment, relying on precise arm motion and detailed models of frictional contact, was formulated. Motion planning for object pushing by computing the set of feasible motions has also been studied [11]. For pivot primitives, on the other hand, a general framework of planning and execution for pivoting with dynamic arm motions was proposed [12]. Q-learning-based open loop pivoting was performed using the inertial forces generated from the robot arm's motion [13]. A forceful manipulation was also presented in a planning approach with a consideration of torque and frictional limits [14]. For practical applications, however, dexterous hands face limitations due to their complex hardware and high cost. Conversely, in-hand manipulation using a simple gripper with external sources can offer a more practical and appropriate solution.

Motion planning approaches for pushing primitives depend on precise robot arm motions and accurate friction models. However, they encounter limitations in real-world scenarios where uncertainties lead to imprecise motion control, or variations in the gripper's tooltip condition and the object's surface state cause changes in the friction model, making it difficult to achieve desired planning results. Therefore, we focus on a real-time force control approach to realize precise placement with pushing primitives in a practical manufacturing scenario such as inserting parts into a machine slot. Control-based in-hand manipulation has been studied to realize various primitives, such as pushing and pivoting. A convex hybrid MPC framework for planar manipulation was proposed, addressing contact mode selection complexity with offline learning [15]. In-hand sliding manipulation with dynamic hand motion was implemented with a lightweight passive gripper [16]. The pivoting of an object was achieved by using gravity and adaptively controlling the gripping forces through force/tactile feedback [17]. A deep predictive model that uses tactile sensor information for active slip control has also been presented [18]. With gravity as an extrinsic resource, a sliding mode controller for in-hand manipulation with visual and tactile feedback was formulated [19]. Swing-up regrasping has been implemented by means of dynamic arm motions based on the energy control approach [20]. Non-prehensile pivoting utilizing contacts from the environment was also implemented with hybrid force–velocity control [21]. Most control-based in-hand manipulations have been realized by visual or/and tactile feedback or a parallel-jaw gripper customization.

The objective of this paper is the formulation and implementation of Cartesian-force-control-based pushing primitives as shown in Figure 1. A hybrid position/force control strategy is suitable for implementing pushing manipulation by controlling the force in a specific Cartesian direction while using position control for the remaining axes. This Cartesian-space hybrid position/force control technique allows the selective application of position or force control to each Cartesian axis by employing a dedicated selection matrix [27]. Careful specification of the selection matrix is critical, as it ensures the commanded motions and forces are consistent with environmental constraints, preventing control errors or instability [28]. This control technique is widely used in various manipulations, such as peg-in-hole assembly [29] or drawing applications [30], where force control is required only along specific axes. We exploit an external environment as a pusher that enables the successful placement of an object in a machine slot with a desired pose regardless of its initial picking pose; the desired configuration of the gripper is for it to be positioned at the edge of the grasped object. One of the key features for such a specific objective is the fast placement of the object. Pose checking with visual information or additional regrasping manipulation are not necessary. To implement Cartesian force control for a robot manipulator, only joint torque information is required, which can be obtained through joint torque sensors or the current signal of joint motors. Motion planning approaches and control-based approaches are compared with our proposed Cartesian-force-control-based pushing primitives, as summarized in Table 1.

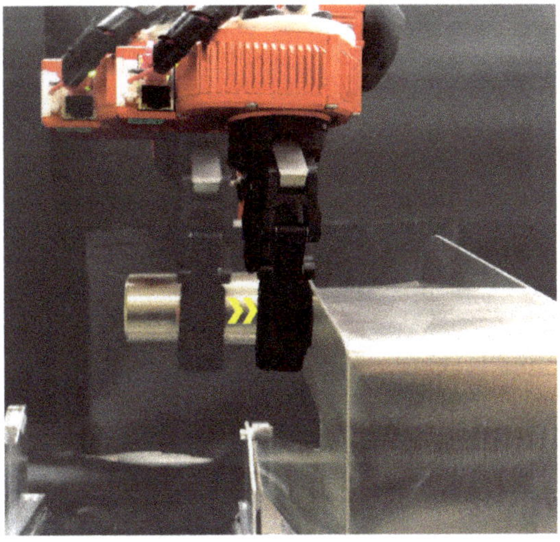

Figure 1. Cartesian-force-control-based pushing primitives.

By controlling the Cartesian force so that it merely exceeds the static friction force between the gripper and the object, sliding motion can be induced between them. Furthermore, the controller ensures that when the gripper makes contact with the external environment, the contact force remains stable by maintaining a consistent Cartesian force. The exact estimation of the frictional contact force between the fingertip and the grasped object is not required. The threshold of the friction force can be determined experimentally. The manufacturing environment, such as the flat surface of a table or a machine fixed on the ground, can be an extrinsic resource for pushing the grasped object. A one-degree-of-freedom (DOF) two-finger gripper is used to realize the pushing action. In addition, only the joint torque sensors of the robot manipulators are used to estimate the Cartesian force feedback. Since we implement Cartesian force control within the robot manipulator instead of grasping force control within the gripper, a tactile sensor at the tooltip to estimate force and torque is not required. Pose feedback sensors such as a camera outside the robot are also not required.

Table 1. Comparison of planning and control methods for in-hand manipulation.

	MIT [9]	Northwestern [16]	ASU [18]	KTH [19]	KIMM
Manipulation	Pushing	Sliding	Pivoting	Pivoting	Pushing
Extrinsic Resources	External pusher	Acceleration	Centrifugal force	Gravity	External pusher
Method	Motion planning	Motion planning	Gripper force control	Gripper force control	Cartesian force control
Feedback	-	-	Tactile	Tactile and Vision	Joint torque sensor
Results	Translational slip	Sliding	Rotational slip	Rotational slip	Translational slip
Requirements	Exact contact (friction) model		Additive feedback information (Vision, Tactile)		-

The main contributions of this work are twofold and can be summarized as follows:

1. Precise placement is achieved using Cartesian-force-control-based pushing primitives, which leverage the surrounding environment and eliminate the need for tactile or vision sensors.
2. The automation of machining processes, traditionally performed manually by humans, is achieved using the proposed approach, enabling the consistent loading of objects into machine slots, and validated on a real testbed.

The rest of this paper is organized as follows. In Section 2, a brief explanation of the target task is provided. The proposed pushing primitives with Cartesian force control are presented in Section 3. An experimental evaluation of the mock-up and the real testbed is presented in Section 4, and conclusive remarks are provided in Section 5.

2. Task Description

We target the automation of the boring machine process used for machining the inner diameter of bushings, a component in automotive manufacturing. In the target process, a worker operates six machines; the scope of the process is to (i) check whether one of the machines needs to be supplied with the bushings to be machined, (ii) grasp the bushings from the bin, (iii) transfer them to the machine, (iv) place the bushings in the machine slot, (v) grasp finished bushings and place them on a tray with an aligned pose, and (vi) repeat the process. Although the process seems to be very simple for a worker, it is rather challenging from the perspective of robot manipulation. In this paper, we focus on the process from (ii) to (iv); these steps are represented in Figure 2. One of the key issues that we are interested in is placing the bushings in the slot with a precise desired posture.

Figure 2. Entire process of the target scenario. The scope of this paper ranges from the grasping to the placing of bushings into a machine slot.

The size and weight of a bushing are 35$\phi \times$L80 mm and 260 g, respectively. The bushing is to be placed in the machine slot in a horizontal posture, and the clearance of the slot is only 5mm, as shown in Figure 3. Therefore, placing a bushing in the slot is difficult if its posture changes even slightly; a worker uses visual information and dexterous hands, which enables complicated manipulation with a combination of pivoting and pushing primitives, and also inserts his finger into the bushing holes to grasp more than one bushing at once. In the case of a robot, although it is difficult to grasp multiple objects simultaneously due to the limited dexterity of the 1-DOF two-finger gripper, simplified manipulation can be achieved through a pushing action that consistently positions the gripper at the edge of the grasped object before placing it into the slot.

Figure 3. Machine slot with 5 mm clearance.

We utilize the external environment as a pusher to facilitate successful in-hand manipulation. A rigid and flat surface, such as a tray, machine surface, or table, can function as a pusher. We selected the left side of the machine slot as an external pusher; the surface is flat and sufficiently rigid, as illustrated in Figure 4.

It is important to note that we initially focused on the bushing machining process to evaluate the feasibility of automating a specific industrial process commonly performed in small and medium-sized enterprises, rather than aiming for generalizability across various scenarios. By demonstrating the feasibility in this particular process, we anticipate that the approach can be extended to a wider range of objects and processes [31].

Figure 4. The left side of the slot can be used as an external pusher.

3. Pushing Primitives with Cartesian Force Control

The pushing action can be realized by exerting Cartesian force control in a direction orthogonal to the grasping direction so that a slip between the gripper and the grasped object can occur. The specific objective is to position the gripper at the edge of the grasped object and this can be achieved by controlling the amount of force used so that is sufficient to make the object slip and is compliant to the environment.

The equation of the motion of a robot manipulator in joint space is as follows [32]:

$$M(q)\ddot{q} + C(q,\dot{q}) + G(q) = \tau - \tau_{ext} \tag{1}$$

where $q \in \mathbf{R}^{n\times 1}$ is the joint position; $M(q) \in \mathbf{R}^{n\times n}$ is the inertia matrix; $C(q,\dot{q}) \in \mathbf{R}^n$ is the vector of centrifugal and Coriolis effects; $G(q)$ is the gravitational torque; $\tau \in \mathbf{R}^n$ is the joint torque; and $\tau_{ext} \in \mathbf{R}^n$ is the external torque exerted on the robot. In this paper, the external torques are measured by joint torque sensors. Let $x(t) \in \mathbf{R}^{6\times 1}$ denote a Cartesian space end-effector position and orientation. The transformation between joint and Cartesian velocities and accelerations can be computed with the Jacobian $J(q) \in \mathbf{R}^{6\times n}$ as follows:

$$\dot{x}(t) = J(q)\dot{q}(t) \tag{2}$$

$$\ddot{x}(t) = \dot{J}(q)\dot{q}(t) + J(q)\ddot{q}(t) \tag{3}$$

Notice that only a nonsingular case is considered in this paper, with the assumption that the Jacobian has a full row rank. The relationship between joint torques and Cartesian forces at the end-effector is given by

$$\tau = J^T(q)F \tag{4}$$

From Equations (2)–(4), the end-effector equation of motion in Cartesian space can be derived in the form [33]

$$\Lambda(x)\ddot{x} + \mu(x,\dot{x}) + p(x) = F - F_{ext} \tag{5}$$

where

$\Lambda(x) = J^{-T}(q)M(q)J^{-1}(q)$;
$\mu(x,\dot{x}) = J^{-T}(q)C(q,\dot{q}) - \Lambda(q)\dot{J}(q)\dot{q}$;
$p(x) = J^{-T}(q)G(q)$.

To globally linearize the robot manipulator dynamics, the computed torque controller in Cartesian space can be formulated as follows [34]:

$$\boldsymbol{F}_c = \boldsymbol{\Lambda}(\boldsymbol{x})\ddot{\boldsymbol{x}}_c + \boldsymbol{\mu}(\boldsymbol{x}, \dot{\boldsymbol{x}}) + \boldsymbol{p}(\boldsymbol{x}) + \boldsymbol{F}_{ext} \tag{6}$$

where $\ddot{x}_c$ can be both the linear position and the force controller. The computed torque controller provides a fundamental framework for designing controllers. It enables the application of various controllers without being constrained by the limitations of nonlinear behavior. The resultant joint torque input can be obtained by Equation (4), $\boldsymbol{\tau}_c = \boldsymbol{J}^T\boldsymbol{F}_c$.

Given that the pushing action is normally performed in a single direction, a hybrid position/force control strategy [27] is suitable for controlling the force for a single pushing direction and position for the rest of the axes. Admittance [35] or impedance [36] controllers could be considered for interaction with the surrounding environment. However, since the goal of implementing the pushing primitive is not to precisely control the amount of slip within the gripper, we determined that implementing Cartesian force control along the desired axis aligns better with our intended approach.

First, the position controller, $\ddot{x}_c = \ddot{x}_p$, can be designed as follows:

$$\ddot{\boldsymbol{x}}_p = \ddot{\boldsymbol{x}}_d + \boldsymbol{K}_{p,d}\dot{\boldsymbol{e}}_p + \boldsymbol{K}_{p,p}\boldsymbol{e}_p \tag{7}$$

where $\boldsymbol{e}_p = \boldsymbol{x}_d - \boldsymbol{x}$ and $\boldsymbol{K}_{p,p}$ and $\boldsymbol{K}_{p,d}$ are the proportional and derivative gain of the position controller. In the position control state, the external force $\boldsymbol{F}_{ext}$ in Equations (5) and (6) is ignored due to an assumption of no presence of contact with the environment. The error dynamics can be obtained as follows:

$$\ddot{\boldsymbol{e}}_p + \boldsymbol{K}_{p,d}\dot{\boldsymbol{e}}_p + \boldsymbol{K}_{p,p}\boldsymbol{e}_p = \boldsymbol{0} \tag{8}$$

The error system is asymptotically stable as long as $\boldsymbol{K}_d$ and $\boldsymbol{K}_p$ are all positive. Second, the force controller, $\ddot{\boldsymbol{x}}_c = \ddot{\boldsymbol{x}}_f$, can be designed as follows:

$$\ddot{\boldsymbol{x}}_f = \ddot{\boldsymbol{F}}_d + \boldsymbol{K}_{f,d}\dot{\boldsymbol{e}}_f + \boldsymbol{K}_{f,p}\boldsymbol{e}_f \tag{9}$$

where $\boldsymbol{e}_f = \boldsymbol{F}_d - \boldsymbol{F}_{ext}$ and $\boldsymbol{K}_{f,p}$ and $\boldsymbol{K}_{f,d}$ are the proportional and derivative gain, respectively, of the force controller. The external force is considered to be in the force control state. The error dynamics are also stable with positive gain, as follows:

$$\ddot{\boldsymbol{e}}_f + \boldsymbol{K}_{f,d}\dot{\boldsymbol{e}}_f + \boldsymbol{K}_{f,p}\boldsymbol{e}_f = 0 \tag{10}$$

A block diagram of the force-controlled pushing action is presented in Figure 5. The $\ddot{\boldsymbol{F}}_d$ and $\dot{\boldsymbol{F}}_d$ terms are neglected because the pushing action needs only constant force to make a slip. The position and force controllers are merged by the policy of a compliance selection matrix that specifies which degrees of freedom are under force control. The components of the selection matrix $\boldsymbol{S} \in \mathbf{R}^{6\times6}$ are as follows:

$$\boldsymbol{S} = \begin{bmatrix} s_1 & & 0 \\ & \ddots & \\ 0 & & s_6 \end{bmatrix} \tag{11}$$

where s_i, $i = 1 \cdots 6$ are selection values that enable the force controller when the value is 1 and the position controller when the value is 0. In the whole process of performing object grasping and placing in the machine slot, we set the force controller up so that it is only enabled for the pushing period. Thus, the force control trigger is introduced; the components of the selection matrix s_i are not fixed and depend on whether the estimated external forces exceed the trigger signal ($s_i = 1$) or not ($s_i = 0$). For the force control trigger in the pushing period, the experimentally determined threshold is used, and for the rest of the period, arbitrary large values are used so as to not enable the force controller. The value of the threshold is used as the desired force in the force control state. The reason for using force control instead of position control during pushing actions is that two undesirable situations can arise because of inaccurate mathematical models, such as friction and robot dynamics, or imprecise motion planning and control; one is deficient pushing due to an unexpectedly large frictional force, and the other is excessive pushing, which may cause a large contact force between the robot and the environment. In reality, the surface friction of the bushings is not uniform due to the presence of oil and metal shavings.

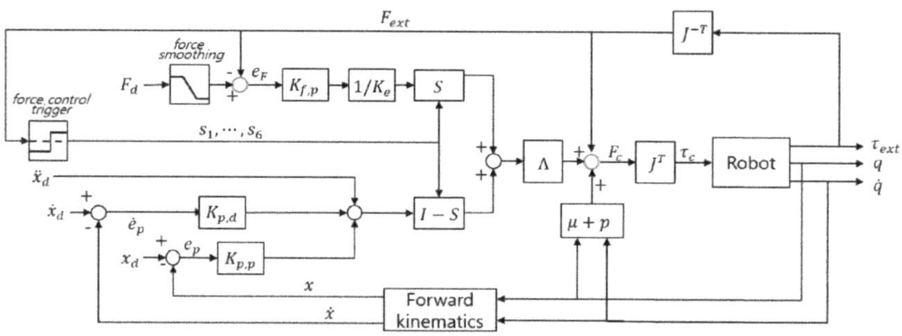

Figure 5. Hybrid position/force control scheme used to realize a Cartesian-force-control-based pushing action. The force control trigger is used to change the control strategy only for the pushing period, and force smoothing is used for smooth control state change.

After the pushing period, the robot manipulator moves to the next point by switching to position control. At this time, because a nonzero force has been generated between the robot and the environment during pushing, a sudden large feedback input may be generated in order to compensate for the contact force and move to the next desired point; thereby, an undesirable recollision with the environment can occur. Therefore, force smoothing is exploited when the control strategy changes from force to position control. The contact force can be diminished to zero gradually by controlling the force necessary to track the diminishing reference force, as follows:

$$\ddot{x}_f(t) = \boldsymbol{K}_{f,p}\Big(\boldsymbol{F}_d(t) \times S_f(t) - \boldsymbol{F}_{ext}(t)\Big) \tag{12}$$

where $S_f(t) \in \mathbf{R}^1$ is a smoothing function multiplied elementwise to each degree of freedom of the reference force. The smoothing function is designed so that, at the start of force control, it takes a step form to induce slip, while at the end of force control, it transitions to a ramp form to gradually reduce the interaction force exerted on the external environment.

4. Experimental Evaluation

Experiments were performed with a simple 1-DOF wire-driven gripper with 2 rigid fingers with common fingertips. The 6-DOF robotic arm kit of HEBI robotics consists of 6 individual Series Elastic Actuators (SEAs); therefore, joint torques can be accurately measured using the strain of a spring internally placed at the actuator output [37]. It is notable that although this study used a serial robotic arm to implement the proposed pick-and-place approach, a parallel-type robot could also be effectively utilized [38]. The proportional and derivative gains of the position controller were set to $K_{p,p} = [300,\ 300,\ 400,\ 20,\ 20,\ 10]$ and $K_{p,d} = [0.1,\ 0.1,\ 0.1,\ 0.1,\ 0.1,\ 0.1]$, respectively. For the force controller, the proportional gains were $K_{f,p} = [0,\ 0.2,\ 0,\ 0,\ 0,\ 0]$, with a nonzero gain applied exclusively to the pushing axis (the y-axis in our configuration).

4.1. Mock-Up Tests

The mock-up hardware represents the bushing machining process, as shown in Figure 6. The machine slot is placed in front of the robot manipulator, and the clearance of the slot is 5 mm, which allows for placement only when the bushing is well positioned. In practice, the bushing is supposed to lie in the bin randomly. However, in this paper, to focus on the problem of placing the bushing in the desired posture, we used a fixed jig to adjust the bushing grasping position arbitrarily to exclude the posture recognition problem (see Figure 7). Although not included in this paper, well-known artificial intelligence (AI)-based algorithms, such as the region-based convolutional neural network (R-CNN), could be possibly useful methods for recognizing the posture of the target bushing.

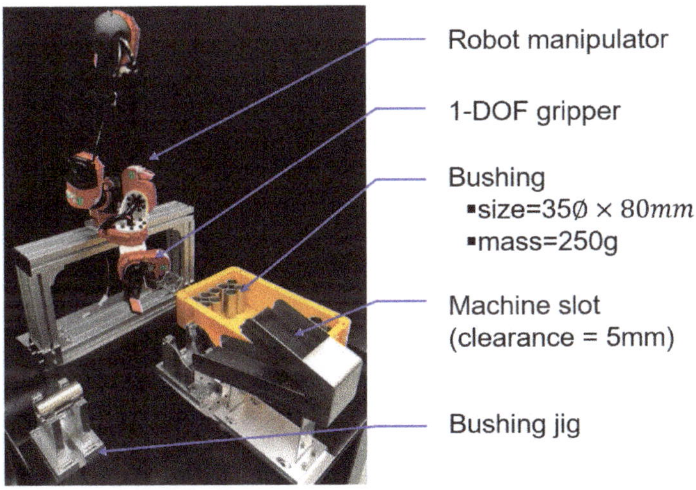

Figure 6. Experimental configuration established to demonstrate real process.

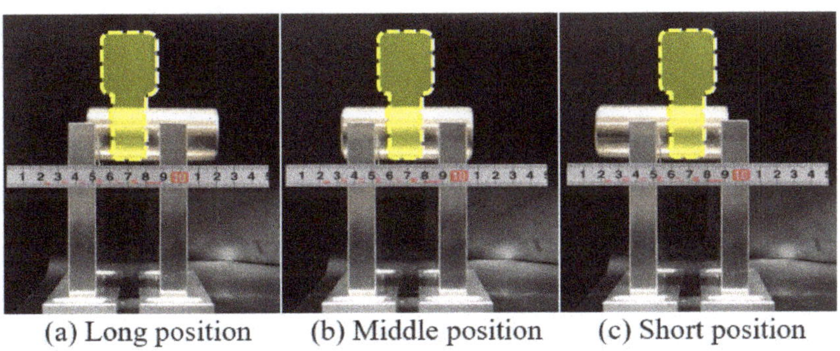

Figure 7. Various grasping positions can be manually adjusted with a jig to express imprecise grasping situations (the yellow dashed lines represent the gripper).

Figure 8 shows the sequences of grasping and placing scenarios that use a Cartesian-force-control-based pushing action. The initial grasping position is manually adjusted to the long and short position of the bushing using a jig. The gripper is located at the end of the bushing in both cases so that placing in the slot is achievable. The left side of the slot is adopted as an external pusher; by doing so, the tac time for the next operation, inserting the bushing into the machine slot, can be minimized. The graphs Figures 9 and 10 represent the detailed results seen during the period from before pushing to finishing. In the pushing period, the position reference is set to a point through the slot in the Y direction. The robot arm begins to approach the external environment (the left side of the slot) with a straight-line trajectory (+Y direction in global coordinates) at 6.3 s (indicated by black dotted vertical line). The entire control input is significantly increased from 7.3 s to 7.7 s because the position controller cannot reduce the position error, which occurs due to contact between the bushing and the slot; the third column of Figure 8 shows this situation. The external contact force is measured by joint torque sensing and the inverse of the Jacobian transpose, $\boldsymbol{F}_e = \boldsymbol{J}^{-T}\boldsymbol{\tau}_e$, as shown in Figure 10. When the external force exceeds the threshold, the force control trigger is enabled, and the control state is changed to the force controller. The desired force of the controller, which is the same magnitude as the trigger threshold, is set to 13 N experimentally to consistently make slip occur between the gripper and the bushing. The position control input disappears, and the force controller maintains the contact force. The aim is to position the gripper at the end of the bushing and there is a fixed pusher at that point, so the magnitude of the force can be quite robust. The behavior of the position control input and the force control input, as shown in Figure 9, represent the characteristics of the smoothing function. To ensure a clear slip between the gripper and the object at the beginning of the pushing phase, the position control input disappears in a step-wise manner, while the force control input is applied in a step-wise fashion. As seen throughout the entire control input, unstable phenomena such as discontinuities rarely occur, and it can be observed that pushing proceeds stably through force control. Conversely, when pushing ends, the force control input decreases in a ramp shape, as shown in Figure 9. Once the force control input completely disappears, the position control input takes over and performs the next operation.

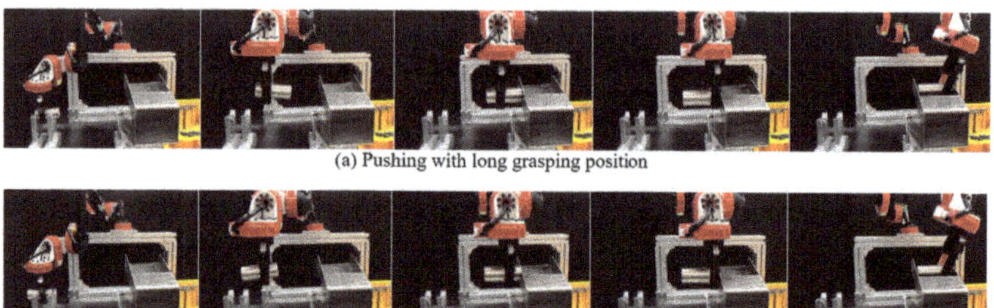

(a) Pushing with long grasping position

(b) Pushing with short grasping position

Figure 8. Experimental sequences from imprecise grasping to precise placing into a slot. The two different grasping positions represented are the long grasping position in the top row and the short grasping position in the bottom row. The snapshots were manually chosen to show the highlighted moments.

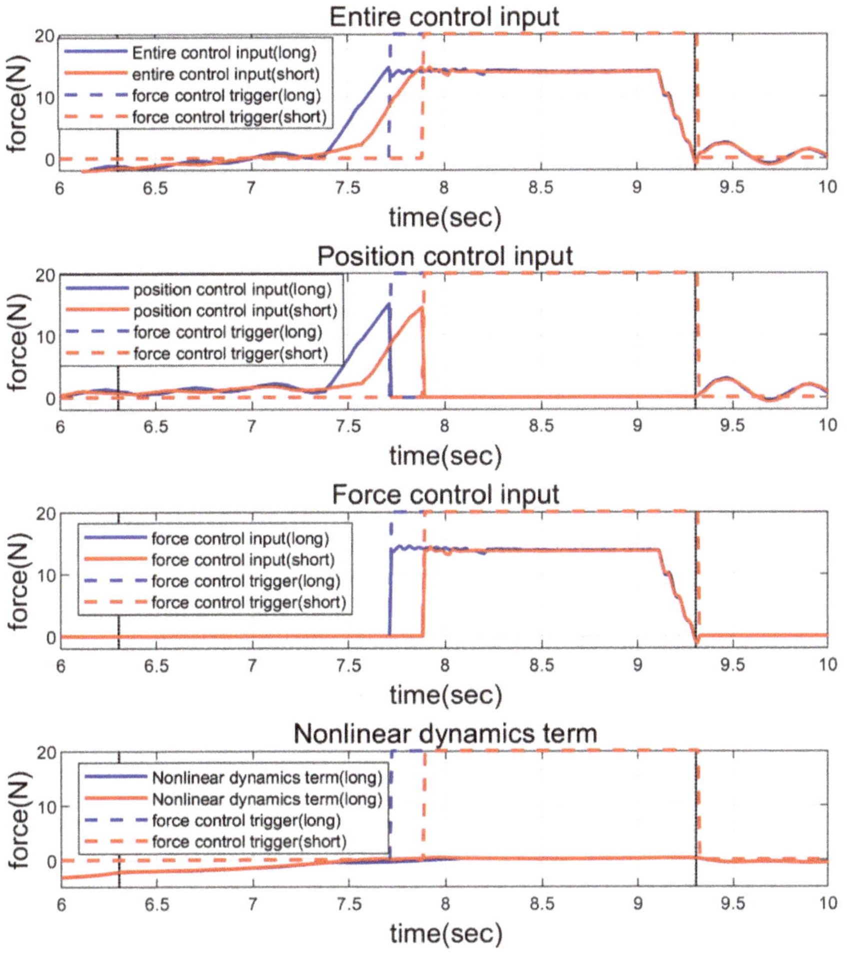

Figure 9. Hybrid position/force control input in the Y direction during the pushing action. The first row is the entire control force, which consists of the position and force control input and the nonlinear dynamics compensation term; the measured external force is included in the force control input. The control strategy is changed when the force control trigger arises.

It is important to note that the friction force may fluctuate due to changes in the condition of the gripper's fingertips or the object's surface, such as when the fingertips become worn, leading to a reduction in frictional force. However, our proposed push manipulation method does not rely on achieving a precise amount of slip. Instead, it requires the desired force to exceed the friction force, but not excessively, to maintain stability. While fine-tuning of the desired force is necessary when the object changes, our approach is not overly sensitive to fluctuations in friction caused by environmental changes. Although not addressed in this paper, adaptive [39] or data-driven [40] control techniques for accurately estimating friction force during pushing manipulation could provide a robust solution to accommodate significant variations in the friction model.

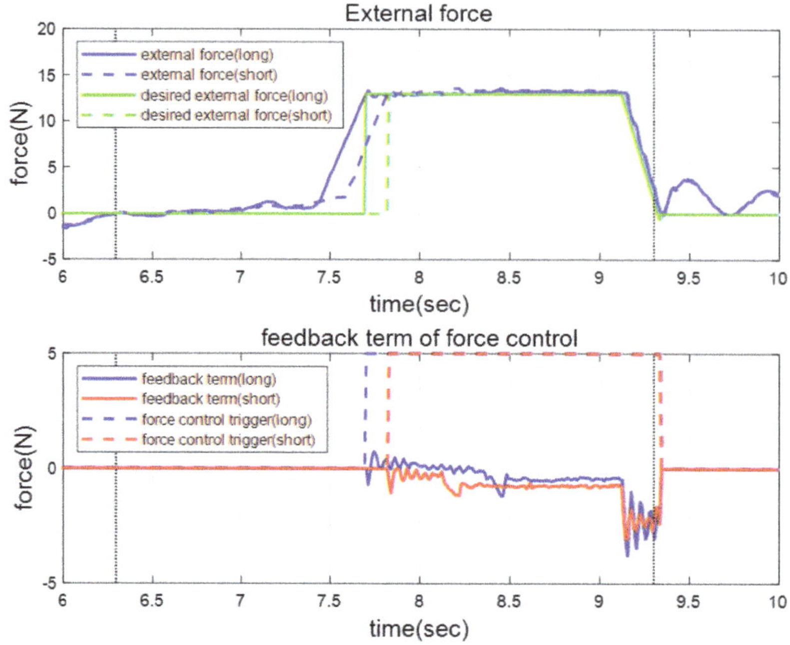

Figure 10. The force control input (Y direction) consists of the measured external force and feedback control term. When the force control is triggered, the desired force is calibrated to induce slipping for the pushing action. The feedback term fluctuates when slip occurs because of uneven contact friction.

The rest of the direction is still controlled by position control, which maintains the pushing pose. The feedback term of the force control contributes to canceling the effect of the uneven frictional contact force. Note that the feedback term in Figure 10 includes vibrations caused by the dynamics of the SEAs as well as fluctuating forces to compensate for uneven friction. To finish the pushing action, the contact force should be diminished to move on to the next point without a sudden large feedback position control input, which may generate an undesirable recollision with the environment. Therefore, force smoothing is exploited at the end of the pushing period. The magnitude of the force reference is decreased from 13 N to 0 N during the smoothing period T_D by the first-order function that follows:

$$\begin{cases} S_f(t) = 1 & for\ T_P - t > T_D \\ S_f(t) = T_D^{-1} \times (T_p - t) & for\ T_P - t \leqslant T_D \end{cases} \tag{13}$$

where T_P is the pushing period and T_D is set to 0.2 s in this paper. The tracking control input is generated during this period.

The time at which the control state change occurs is compared for the two cases of long and short grasping positions. The time at which this change occurs for the long grasping position is twice longer than that of the short grasping position. The time difference implies a different initial grasping position; in this case, the long grasping position is twice the length of the short grasping position. The end of the slip motion can be inferred from the end of fluctuation in the feedback term signal.

4.2. Application to Testbed

A real-scale boring machine testbed was constructed to evaluate the practicality of the proposed pushing-based control method, as illustrated in Figure 11. The overall setup of the boring machine testbed and the bushing placement process are presented. The setup specifically integrates a robot manipulator and external elements, such as a worktable and slots, enabling pushing actions that effectively leverage the surrounding environment.

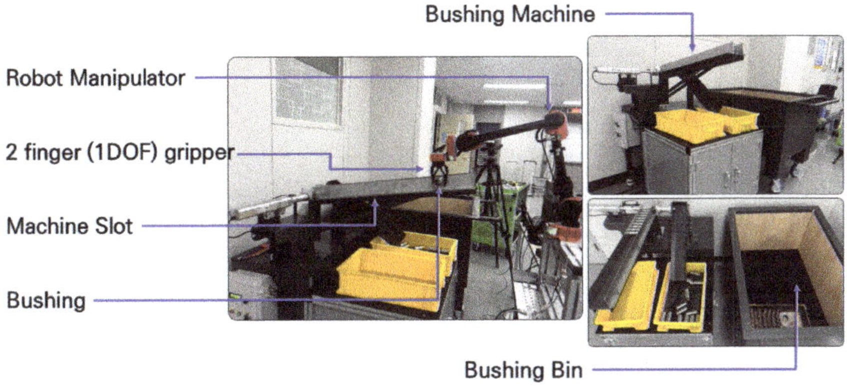

Figure 11. A testbed configuration describing real-world bushing machine as illustrated in Figure 2.

Detailed sequences of the experiments conducted on the testbed, showcasing the pushing process and the insertion of the bushing into the slot, are illustrated in Figure 12. This also demonstrates the robot's ability to successfully place the bushing within the slot with a clearance of approximately 5 mm. Figure 13 highlights the consistent placement of the bushing into the slot, irrespective of its initial position, further validating the feasibility of the proposed method. The entire scenario is available at https://www.youtube.com/watch?v=TSW7lbNnbRQ (accessed on 2 December 2024), where three consecutive operations were captured in a single continuous take.

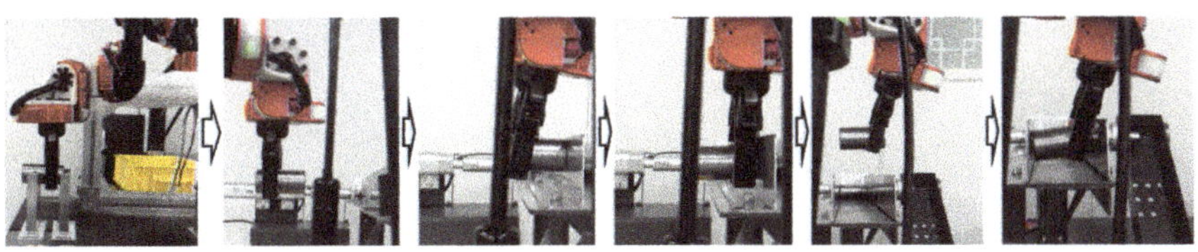

Figure 12. Detailed experimental results from the testbed, illustrating the pushing process and the insertion of the bushing into the slot.

The proposed method was compared to two alternative approaches. The first approach attempts to place the bushing into the machine slot without considering the initial position, while the second approach implements pushing primitives that use motion planning instead of a control-based method. The method that does not consider the initial position only succeeded when the gripper precisely grasped the bushing at a specific point. Our experiments conducted with three initial positions, short, middle, and long, resulted in a success rate of 33%. In practice, the slot's clearance is limited to approximately 5 mm, allowing successful placement only when the grasping position falls within this range. The motion planning approach also showed a low success rate. This method requires predefined slip distances between the gripper and the bushing. Without additional vision sensors for real-time feedback of the initial grasp position, it can only handle one of the three initial positions. Its success rate was limited to 27%, including failures caused by inconsistencies in the grasp friction state, which prevented accurate implementation of the desired slip distance. Moreover, attempting to create more slip than necessary to reach the edge increased the contact force between the gripper and the external pusher, causing the manipulator to diverge. This instability makes the motion planning approach unreliable. In contrast, the proposed control-based method maintained a consistent force during slip, allowing the gripper to position the bushing at the edge accurately, regardless of its initial position. The method achieved a 97% success rate, with the only failure case occurring when the bushing slipped out of the gripper during slip control. Importantly, the proposed method does not cause divergence or other issues even in cases of failure, making it a stable option for process automation. The results of the comparison among the three methods are summarized in Table 2.

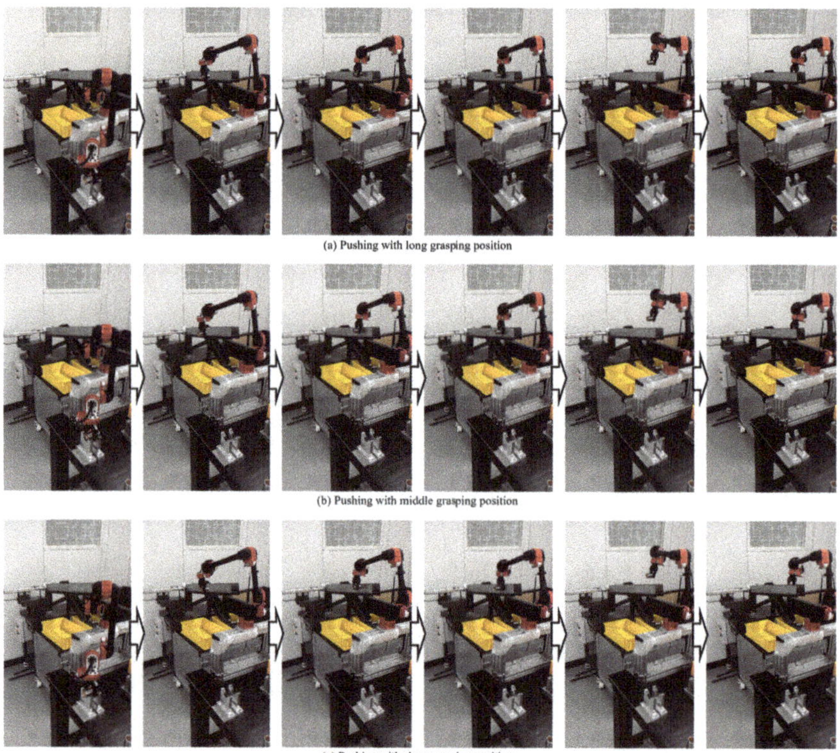

Figure 13. Experimental results from the testbed showing successful insertions into the slot by the precise placement of the bushing when the gripper is positioned at (**a**) short, (**b**) middle, and (**c**) long initial grasping positions.

Table 2. Comparison of success rates.

Method	Success Rate (%)	Success/Attempts
No-Pushing Approach	33	10/30
Motion Planning Pushing Approach	27	9/30
Proposed Controlled Pushing Approach	97	29/30

It should be noted that both the no-pushing approach and the motion planning pushing approach are fundamentally constrained to a maximum success rate of 33% due to the lack of grasp position information. In contrast, the proposed controlled pushing approach significantly improves the success rate by addressing practical issues, such as maintaining the fingertip condition to prevent changes in frictional force. Additionally, as represented in Figure 13, the use of a simple two-finger gripper introduces inevitable changes in the gripping angle after pushing manipulation, which are accommodated for by adapting the scenario to ensure successful slot insertion.

To summarize, in situations where the accuracy of the initial grasp position cannot be guaranteed, or real-time visual feedback of the grasp state is unavailable, the proposed control-based pushing primitive proves significantly more effective than the motion planning approach. It is noted that the robot's operating efficiency is noticeably slower than that of humans. Despite this, the validity of the proposed bushing machining process lies in its achievement of automation of the target industrial process by consistently supplying the buffer without depletion.

5. Conclusions

In-hand manipulation with force-control-based pushing primitives is formulated and implemented to achieve the precise placement of bushings in a machine slot, even under conditions that make it difficult to grasp from a planned position. To avoid difficulty in the mathematical modeling of friction between the fingertips and grasped objects, we adopted a strategy that places the gripper at the edge of the grasped object by exploiting prehensile pushing actions using an external environment, regardless of the initial grasping position. Therefore, detecting and maintaining the contact force between the gripper and the external environment is of primary importance and can be accomplished by force feedback control. As external forces are estimated by joint torque sensing, additional sensors, such as vision or tactile sensors, are not necessary.

A smoothing function was introduced to effectively create slip between the fingertip and the object at the start of pushing that used a step-based form of interaction force, and to smoothly conclude the interaction between the gripper and the external pusher at the end, a ramp-based form of interaction force was used. The experimental results demonstrated the feasibility of the proposed method, showing that the bushing object could be precisely placed into the machine slot despite variations in initial grasping positions. Compared to other methods, such as the no-pushing approach and the planning pushing approach, which achieved only a 33% success rate, our proposed method achieved a significantly higher success rate of 97%.

Our proposed method leverages a feedback loop and an external pusher, which implies that it does not require an accurate friction model. More importantly, this approach allows the force controller for the pushing axis to be implemented using traditional PID or LQ control techniques as well as heuristic approaches [40,41]. As a result, our approach demonstrates the feasibility of automating industrial processes without the need for complex control methodologies, which is a significant advantage.

The authors' ongoing research is into integrating the grasping process, which includes recognizing an object's pose by utilizing a vision sensor fixed at the wrist of the manipulator, and expanding their research to various types of objects. Data-driven techniques for estimating frictional force during pushing manipulation will also be considered to address various uncertainties, such as surface inconsistencies or the presence of debris, and maximize automation.

Author Contributions: Conceptualization, J.P., J.-J.K., and D.-Y.K.; methodology, J.P. and J.-J.K.; software, J.P. and J.-J.K.; validation, J.P. and D.-Y.K.; formal analysis, J.P., J.-J.K., and D.-Y.K.; investigation, J.P.; resources, J.P.; data curation, J.-J.K. and D.-Y.K.; writing original draft preparation, J.P.; writing review and editing, J.-J.K. and D.-Y.K.; visualization, J.P.; supervision, J.P.; project administration, J.-J.K.; funding acquisition, J.P. All authors have read and agreed to the published version of the manuscript.

Funding: This research was supported by the National Research Council of Science & Technology as part of the project entitled "Developed core technologies for the Robotic General Purpose Task Artificial Intelligence (RoGeTA) framework to enable a variety of everyday services" (NK248G) and the Industrial Strategic Technology Development Program (no. 20018745 and no. 00416440) funded by the Ministry of Trade, Industry & Energy (MOTIE, Korea).

Data Availability Statement: The data presented in this study are available on request from the corresponding author.

Conflicts of Interest: The authors declare no conflicts of interest.

References

1. Monkman, G.J.; Hesse, S.; Steinmann, R.; Schunk, H. *Robot Grippers*; John Wiley & Sons: Hoboken, NJ, USA, 2007.
2. Morino, K.; Kikuchi, S.; Chikagawa, S.; Izumi, M.; Watanabe, T. Sheet-Based Gripper Featuring Passive Pull-In Functionality for Bin Picking and for Picking Up Thin Flexible Objects. *IEEE Robot. Autom. Lett.* **2020**, *5*, 2007–2014. [CrossRef]
3. Zhao, C.; Tong, Z.; Rojas, J.; Seo, J. Learning to Pick by Digging: Data-Driven Dig-Grasping for Bin Picking from Clutter. In Proceedings of the 2022 IEEE International Conference on Robotics and Automation (ICRA), Philadelphia, PA, USA, 23–27 May 2022; pp. 749–754.
4. Lévesque, F.; Sauvet, B.; Cardou, P.; Gosselin, C. A model-based scooping grasp for the autonomous picking of unknown objects with a two-fingered gripper. *Robot. Auton. Syst.* **2018**, *106*, 14–253.
5. Babin, V.; Gosselin, C. Picking, grasping, or scooping small objects lying on flat surfaces: A design approach. *Int. J. Robot. Res.* **2018**, *37*, 1484–1499. [CrossRef]
6. Zeng, A.; Song, S.; Welker, S.; Lee, J.; Rodriguez, A.; Funkhouser, T. Learning Synergies Between Pushing and Grasping with Self-Supervised Deep Reinforcement Learning. In Proceedings of the 2018 IEEE/RSJ International Conference on Intelligent Robots and Systems (IROS), Madrid, Spain, 1–5 October 2018; pp. 4238–4245.
7. Lee, J.; Cho, Y.; Nam, C.; Park, J.; Kim, C. Efficient Obstacle Rearrangement for Object Manipulation Tasks in Cluttered Environment. In Proceedings of the 2019 IEEE International Conference on Robotics and Automation (ICRA), Montreal, QC, Canada, 20–24 May 2019; pp. 183–189.
8. Berscheid, L.; Meißner, P.; Kröger, T. Robot Learning of Shifting Objects for Grasping in Cluttered Environments. In Proceedings of the 2019 IEEE/RSJ International Conference on Intelligent Robots and Systems (IROS), Macau, China, 3–8 November 2019; pp. 612–618.
9. Dafle, N.C.; Rodriguez, A.; Paolini, R.; Tang, B.; Srinivasa, S.S.; Erdmann, M.; Mason, M.T.; Lundberg, I.; Staab, H.; Fuhlbrigge, T. Extrinsic dexterity: In-hand manipulation with external forces. In Proceedings of the 2014 IEEE International Conference on Robotics and Automation (ICRA), Hong Kong, China, 31 May–7 June 2014; pp. 1578–1585.
10. Dafle, N.C.; Rodriguez, A. Prehensile pushing: In-hand manipulation with push-primitives. In Proceedings of the 2015 IEEE/RSJ International Conference on Intelligent Robots and Systems (IROS), Hamburg, Germany, 28 September–2 October 2015; pp. 6215–6222.
11. Chavan-Dafle, N.; Holladay, R.; Rodriguez, A. In-Hand Manipulation via Motion Cones. *Robot. Sci. Syst. (Rss)* **2018** .
12. Holladay, A.; Paolini, R.; Mason, M.T. A general framework for open-loop pivoting. In Proceedings of the 2015 IEEE International Conference on Robotics and Automation (ICRA), Seattle, WA, USA, 26–30 May 2015; pp. 3675–3681.
13. Cruciani, S.; Smith, C. In-Hand Manpulation Using Three-Stages Open Loop Pivoting. In Proceedings of the 2017 IEEE/RSJ International Conference on Intelligent Robots and Systems (IROS), Vancouver, BC, Canada, 24–28 September 2017; pp. 1244–1251.
14. Holladay, R.; Lozano-Pérez, T.; Rodriguez, A. Robust planning for multi-stage forceful manipulation. *Int. J. Robot. Res.* **2023**, *43*, 330–353. [CrossRef]
15. Hogan, F.R.; Grau, E.R.; Rodriguez, A. Reactive Planar Manipulation with Convex Hybrid MPC. In Proceedings of the 2018 IEEE International Conference on Robotics and Automation (ICRA), Brisbane, Australia, 21–25 May 2018; pp. 247–253.

16. Shi, J.; Woodruff, J.Z.; Umbanhowar, P.B.; Lynch, K.M. Dynamic in-hand sliding manipulation. *IEEE Trans. Robot.* **2017**, *33*, 778–795. [CrossRef]
17. Costanzo, M.; Maria, G.D.; Na, C. Slipping Control Algorithms for Object Manipulation with Sensorized Parallel Gripper. In Proceedings of the 2018 IEEE International Conference on Robotics and Automation (ICRA), Brisbane, Australia, 21–25 May 2018; pp. 7455–7461.
18. Stepputtis, S.; Yang, Y.; Amor, H.B. Extrinsic Dexterity Through Active Slip Control Using Deep Predictive Models. In Proceedings of the 2018 IEEE International Conference on Robotics and Automation (ICRA), Brisbane, Australia, 21–25 May 2018; pp. 3180–3185.
19. Karayiannidis, F.E.V.B.Y.; Pauwels, K.; Smith, C.; Kragic, D. In-hand manipulation using gravity and controlled slip. In Proceedings of the 2015 IEEE/RSJ International Conference on Intelligent Robots and Systems (IROS), Hamburg, Germany, 28 September–2 October 2015; pp. 5636–5641.
20. Sintov, A.; Sha, A. Swing-up Regrasping Algorithm using Energy. In Proceedings of the 2018 IEEE International Conference on Robotics and Automation (ICRA), Stockholm, Sweden, 16–21 May 2018; pp. 4888–4893.
21. Hou, Y.; Mason, M.T. Robust Execution of Contact-Rich Motion Plans by Hybrid Force-Velocity Control. In Proceedings of the 2019 IEEE International Conference on Robotics and Automation (ICRA), Montreal, QC, Canada, 20–24 May 2019; pp. 1933–1939.
22. Tournassoud, P.; Lozano-Përez, T.; Mazer, E. Regrasping. In Proceedings of the 1987 IEEE International Conference on Robotics and Automation (ICRA), Raleigh, NC, USA, 31 March–3 April 1987; pp. 1924–1928.
23. Lozano-Përez, T.; Jones, J.; Mazer, E.; O'Donnell, P.; Grimson, W.; Tournassoud, P.; Lanusse, A. Handey: A robot system that recognizes, plans, and manipulates. In Proceedings of the 1987 IEEE International Conference on Robotics and Automation (ICRA), Raleigh, NC, USA, 31 March–3 April 1987; pp. 843–849.
24. Salisbury, J.K.; Craig, J.J. Articulated hands force control and kinematic issues. *Int. J. Robot. Res.* **1982**, *1*, 4–17. [CrossRef]
25. Wimbock, T.; Ott, C.; Hirzinger, G. Passivity-based object-level impedance control for a multifingered hand. In Proceedings of the 2006 IEEE/RSJ International Conference on Intelligent Robots and Systems (IROS), Beijing, China, 9–15 October 2006; pp. 4621–4627.
26. Yoshikawa, T. Multifingered robot hands: Control for grasping and manipulation. *Annu. Rev. Control* **2010**, *34*, 199–208. [CrossRef]
27. Raibert, M.H.; Craig, J.J. Hybrid Position/Force Control of Manipulators. *J. Dyn. Syst. Meas. Control* **1981**, *103*, 126–133. [CrossRef]
28. Duffy, J. The fallacy of modern hybrid control theory that is based on "orthogonal complements" of twist and wrench spaces. *J. Field Robot.* **1990**, *7*, 139–144. [CrossRef]
29. Park, H.; Bae, J.; Park, J.; Baeg, M.; Park, J. Intuitive peg-in-hole assembly strategy with a compliant manipulator. In Proceedings of the IEEE ISR 2013, Seoul, Republic of Korea, 24–26 October 2013; pp. 1–5.
30. Ding, S.; Peng, J.; Xin, J.; Zhang, H.; Wang, Y. Task-Oriented Adaptive Position/Force Control for Robotic Systems Under Hybrid Constraints. *IEEE Trans. Ind. Electron.* **2024**, *71*, 12612–12622. [CrossRef]
31. Qi, W.; Xu, X.; Qian, K.; Schuller, B.W.; Fortino, G.; Aliverti, A. A Review of AIoT-based Human Activity Recognition: From Application to Technique. *IEEE J. Biomed. Health Inform.* **2024**, 1–14. [CrossRef] [PubMed]
32. Craig, J.J. *Introduction to Robotics Mechanics and Control*; Pearson Prentice Hall Pearson Education, Inc.: Upper Saddle River, NJ, USA, 2005.
33. Khatib, O. A unified approach for motion and force control of robot manipulators: The operational space formulation. *IEEE J. Robot. Autom.* **1987**, *3*, 43–53. [CrossRef]
34. Lynch, K.M.; Park, F.C. *Modern Robotics*; Cambridge University Press: Cambridge, UK, 2017.
35. Keemink, A.Q.; van der Kooij, H.; Stienen, A.H. Admittance control for physical human–robot interaction. *Int. J. Robot. Res.* **2018**, *37*, 1421–1444. [CrossRef]
36. Park, J.; Shin, Y.; Kim, S. Object-Aware Impedance Control for Human–Robot Collaborative Task With Online Object Parameter Estimation. *IEEE Trans. Autom. Sci. Eng.* **2024**, 1–14. [CrossRef]
37. Rollinson, D.; Bilgen, Y.; Brown, B.; Enner, F.; Ford, S.; Layton, C.; Rembisz, J.; Schwering, M.; Willig, A.; Velagapudi, P.; et al. Design and Analysis of Series Elasticity in Closed-Loop Actuator Force Control. In Proceedings of the 2014 IEEE/RSJ International Conference on Intelligent Robots and Systems (IROS), Chicago, IL, USA, 14–18 September 2014.
38. Wang, D.; Wang, L.; Wu, J.; Ye, H. An Experimental Study on the Dynamics Calibration of a 3-DOF Parallel Tool Head. *IEEE/ASME Trans. Mechatron.* **2019**, *24*, 2931–2941. [CrossRef]
39. Jleilaty, S.; Ammounah, A.; Abdulmalek, G.; Nouvelière, L.; Su, H.; Alfayad, S. Distributed real-time control architecture for electrohydraulic humanoid robots. *Robot. Intell. Autom.* **2024**, *44*, 607–620. [CrossRef]

40. Zhao, J.; Wang, Z.; Lv, Y.; Na, J.; Liu, C.; Zhao, Z. Data-Driven Learning for H ∞ Control of Adaptive Cruise Control Systems. *IEEE Trans. Veh. Technol.* **2024**, *73*, 18348–18362. [CrossRef]
41. Su, H.; Ovur, S.E.; Xu, Z.; Alfayad, S. Exploring the Potential of Fuzzy Sets in Cyborg Enhancement: A Comprehensive Review. *IEEE Trans. Fuzzy Syst.* **2024**, 1–19. [CrossRef]

applied sciences

MDPI

Article

See-Then-Grasp: Object Full 3D Reconstruction via Two-Stage Active Robotic Reconstruction Using Single Manipulator

Youngtaek Hong [1], Jonghyeon Kim [1], Geonho Cha [2], Eunwoo Kim [3,*] and Kyungjae Lee [4,*]

1 Department of Artificial Intelligence, Chung-Ang University, Seoul 06974, Republic of Korea; volunt4s@cau.ac.kr (Y.H.); kjh980610@cau.ac.kr (J.K.)
2 NAVER Cloud, Seongnam 13561, Republic of Korea; geonho.cha@navercorp.com
3 Department of Computer Science and Engineering, Chung-Ang University, Seoul 06974, Republic of Korea
4 Department of Statistics, Korea University, Seoul 02841, Republic of Korea
* Correspondence: eunwoo@cau.ac.kr (E.K.); kyungjae_lee@korea.ac.kr (K.L.)

Abstract: In this paper, we propose an active robotic 3D reconstruction methodology for achieving full object 3D reconstruction. Existing robotic 3D reconstruction approaches often struggle to cover the entire view space of the object or reconstruct occluded regions, such as the bottom or back side. To address these limitations, we introduce a two-stage robotic active 3D reconstruction pipeline, named See-Then-Grasp (STG), that employs a robot manipulator for direct interaction with the object. The manipulator moves toward the points with the highest uncertainty, ensuring efficient data acquisition and rapid reconstruction. Our method expands the view space of the object to include the entire perspective, including occluded areas, making the previous fixed view candidate approach time-consuming for identifying uncertain regions. To overcome this, we propose a gradient-based next best view pose optimization method that efficiently identifies uncertain regions, enabling faster and more effective reconstruction. Our method optimizes the camera pose based on an uncertainty function, allowing it to identify the most uncertain regions in a short time. Through experiments with synthetic objects, we demonstrate that our approach effectively addresses the next best view selection problem, achieving significant improvements in computational efficiency while maintaining high-quality 3D reconstruction. Furthermore, we validate our method on a real robot, showing that it enables full 3D reconstruction of real-world objects.

Keywords: active 3D reconstruction; object full 3D object reconstruction; robot vision

1. Introduction

Robotic manipulation has emerged as a critical area of research, both in industrial applications [1–3] and in everyday home environments [4–6], with significant attention focused on enabling robots to interact effectively with objects in their surroundings. A fundamental challenge in this domain is the need for comprehensive 3D information about objects, such as geometric shapes and structural details. Traditional methods in object manipulation, which rely on pre-existing 3D models, perform well with familiar objects but falter when faced with novel ones. Recent learning-based approaches show promise for simpler tasks with abundant training data but often struggle with more complex tasks, such as novel object assembly and rearrangement. In these cases, the ability to leverage 3D models becomes essential, underscoring the importance of active 3D reconstruction in enhancing robotic capabilities in diverse and unstructured environments.

In 3D modeling, traditional methods based on sensors, such as LiDAR [7,8] or structured light sensors [9,10], have been used for 3D reconstruction of novel objects. These high-cost sensors capture detailed depth information and are capable of reconstructing fine geometry. However, these methods often necessitate changes in various settings and configurations during the reconstruction process, leading to inevitable human intervention and increased complexity.

To overcome the limitations of these high-cost sensors, recent advancements have focused on using RGB cameras for 3D reconstruction. One such method is structure-from-motion (SfM), which also uses RGB images to reconstruct 3D models. However, SfM often requires multiple viewpoints and precise alignment to generate accurate geometry and can be computationally expensive for complex objects. More recently, neural implicit representations, such as neural radiance fields (NeRF) [11], have emerged as an alternative, enabling 3D reconstruction of unknown objects using only RGB images. NeRF has garnered significant attention in the robot vision field due to its ability to produce high-quality 3D reconstructions without requiring high-cost sensors. NeRF's scene-agnostic 3D reconstruction capabilities have facilitated a range of complex robotic downstream tasks [12–16].

Building on these advances, recent studies in active robotic 3D reconstruction have increasingly incorporated neural implicit representations [17–20]. However, most research in this area assumes rooted objects, which prevents the reconstruction of hidden surfaces such as the bottom of an object, the back of a partially occluded object, or intricate details in cavities and recesses. This issue is particularly pronounced in single fixed-based robotic manipulator systems, where the reachable view space is limited to the manipulator's workspace, restricting object 3D reconstruction to these limited viewpoints [21,22]. As a result, critical parts of the object, such as its bottom side or occluded regions, may not be reconstructed accurately, leading to incomplete models.

To address these challenges, we propose a two-stage robotic active 3D reconstruction pipeline, named See-Then-Grasp (STG), using two cameras within a single robot manipulator system. In the first stage, a hand-eye camera captures the object's approximate geometry. In the second stage, the robot grasps the object and presents it to a fixed external camera, achieving full 3D coverage. To expedite viewpoint selection, we introduce a gradient-based pose optimization technique, enabling the robot to efficiently identify and observe uncertain regions, ensuring a complete 3D reconstruction. Our proposed pipeline is illustrated in Figure 1.

Our contributions can be summarized as follows:

- We propose a novel two-stage active robotic 3D reconstruction pipeline to address the limitations of fixed-base single manipulator systems, which are constrained to partial object reconstruction due to restricted view space. Our pipeline leverages two cameras to achieve full 3D reconstruction of objects by combining hand-eye and external fixed camera perspectives.
- We develop a gradient-based next best view selection method, designed to efficiently explore the full view space of an object by identifying uncertain regions. This method leverages uncertainty in neural implicit representations to optimize view poses for uncovering unseen regions.
- Through experiments on synthetic objects, we demonstrate that our method finds uncertain regions faster and more accurately compared to other approaches. Additionally, by applying our method to a robotic system, we successfully enable full 3D reconstruction of novel real-world objects.

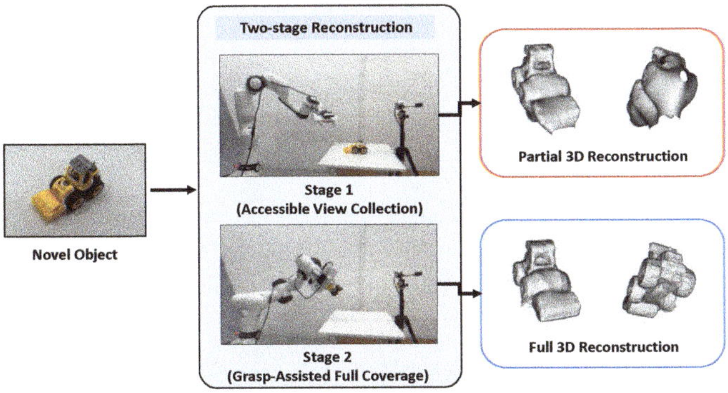

Figure 1. Overview. We propose an active learning framework that automates the full 3D reconstruction of novel objects using a single robot manipulator system. As shown in the red box, a fixed-base single manipulator system can only partially reconstruct the front part of an object due to the constraints of the robot's view space. By utilizing our proposed See-Then-Grasp method, the robot can directly interact with the object, granting access to the entire view space of the object. As illustrated in the blue box, this allows for full 3D reconstruction.

2. Related Work

2.1. Autonomous 3D Reconstruction in Robotic System

Autonomous 3D reconstruction of objects using robots is a long-standing problem. Previous works have employed various robotic platforms for this task, including Unmanned Aerial Vehicles (UAVs) [19,20], wheeled mobile robots [17,23], and fixed-base robotic arms [9,21,22,24].

To model the entire geometry of an object, a robot must have access to the entire view space that allows for the observation of all sides of the object. In this regard, UAVs and wheeled mobile robots are frequently used robotic platforms as they can cover a 360° view of the object without space restriction. Specifically, Lee et al. [17] generate finite-view candidates that cover the whole-object geometry, allowing the mobile robot to access view candidates. Yan et al. [19] use UAVs to freely access the view space within a continuous SE(3) manifold.

In the fixed-based robotic arm setup, the workspace of the robot is limited, resulting in the inability to access the view space of static objects, such as the back or bottom side of the objects. To address this limitation, Wu et al. [9] utilize a multi-robot system to expand the view space of static objects. Lee et al. [24] employ an additional rotary table to enable a single robot system to access the 360° view space of the object.

Most 3D reconstruction approaches in robotic applications often rely on RGB-D data captured by sensors such as Microsoft Kinect, leveraging both color and depth information for high-accuracy reconstructions [25–28]. For instance, KinectFusion [26] uses RGB-D data to perform real-time 3D reconstruction by aligning and merging depth frames. Similarly, Isler et al. [25] formulate volumetric information gain using RGB-D sensors, enabling active reconstruction with mobile robots. However, these methods depend on specialized depth sensors, which can be costly, less scalable, and sensitive to factors like lighting, calibration, and object reflectivity.

In our work, we propose a method that overcomes these limitations by utilizing only RGB cameras combined with neural implicit representation. By directly manipulating objects within a single robotic arm setup, our approach enables complete 3D reconstruction, including occluded regions, without the need for depth sensors, making it more cost-effective and robust.

2.2. Neural Implicit Representation

Neural implicit representations, such as NeRF [11], have gained significant attention in the fields of novel view synthesis [29] and object surface reconstruction [30,31] due to their geometry-agnostic nature and flexible training pipelines. Notably, neural implicit representations enable high-quality 3D scene modeling using only 2D images, prompting numerous efforts to integrate these representations into robotic tasks such as object grasping [15,16], SLAM [14], and visual navigation [12,13].

To apply neural implicit representations in robotic systems, fast and efficient training is crucial. Recent advancements have utilized various techniques to enhance training speeds. Instant-NGP [32] uses hash encoding to provide efficient data encoding while reducing memory usage, allowing for rapid scene modeling. DVGO [33] combines explicit voxel grids with neural networks to enable fast reconstruction of high-resolution 3D scenes. TensoRF [34] leverages high-dimensional tensor decomposition techniques to enhance the expressiveness of NeRF models and improve training efficiency, achieving superior performance even in complex scenes.

Our primary goal is to quickly reconstruct the 3D surface of objects within a robotic framework using sparse object views. To this end, we adopt Voxurf [35], which uses neural SDFs for accurate object surface prediction and voxel grid representation to enable a fast training speed.

2.3. Uncertainty Measurement for Next Best View Selection

A next best view (NBV) selection is a robot vision problem in which the robot chooses the optimal viewpoint to minimize uncertainty in an unseen environment or object model. To solve the NBV selection problem efficiently, a high-quality uncertainty measurement is crucial.

Recently, using neural implicit representation for uncertainty quantification in NBV has been attracting attention. Refs. [36–38] have proposed a stochastic NeRF incorporating Variational Inference [39] into the NeRF learning pipeline. Refs. [40,41] have designed additional MLPs within the NeRF model to directly quantify uncertainty. Ref. [42] utilizes a deep ensemble approach [43] to quantify NeRF uncertainty through the variance of ensemble members.

Although prior works have shown successful results, these approaches require modifications to the NeRF model architecture or involve an increase in the number of network parameters, which extends rendering times. This can lead to increased decision-making times, making them potentially unsuitable for robotic frameworks.

Consequently, we adopt volumetric uncertainty by calculating the entropy of ray weight distribution, as in [17,44]. Using entropy-based uncertainty allows for a focused analysis of an object's 3D geometry and enables rapid uncertainty calculation without the need for additional training or NeRF model modification.

3. Background

Three-dimensional representation with NeRF [11] generally utilizes multilayer perceptrons (MLPs) to model complex 3D scene geometry. Specifically, the MLPs take 3D coordinates $\mathbf{x} = (x, y, z)$ and 2D viewing directions $\mathbf{d} = (\theta, \phi)$ to predict the RGB colors $\mathbf{c}$ and volumetric density σ as

$$\mathrm{MLP}(\mathbf{x}, \mathbf{d}) = (\mathbf{c}, \sigma) : \mathbb{R}^3 \times \mathbb{R}^2 \mapsto \mathbb{R}^4. \tag{1}$$

Then, once the NeRF MLP is trained, $\mathbf{c}$ and σ can be used to synthesize images from novel viewpoints through the volume-rendering method [45]. In the field of robotics, the volumetric density σ can further be used to extract the geometric information of objects, which can then be utilized for manipulation tasks.

In volume rendering, the color of each pixel is rendered as the expected color computed along with the ray from the camera's origin to the pixel location. Since the NeRF dataset consists of pairs of ground truth RGB images $\mathcal{I}$ and camera poses $\mathcal{T} \in \mathrm{SE}(3)$, denoted as $\{\mathcal{I}_i, \mathcal{T}_i\}_{i=1}$, the origin and viewing direction of the camera can be obtained from the corresponding camera pose $\mathcal{T}$. Given the camera's origin $\mathbf{o}$ and viewing direction $\mathbf{d}$ based on the pixel location, the ray is defined as a half-line starting from the origin, i.e., $\mathbf{r}(t) = \mathbf{o} + t\mathbf{d}$ for $t \geq 0$. The expected color $C(\mathbf{r})$ along with the ray $\mathbf{r}$ is computed as

$$C(\mathbf{r}) := \int_0^{\infty} T(t)\sigma(\mathbf{r}(t))\mathbf{c}(\mathbf{r}(t), \mathbf{d})dt, \tag{2}$$

$$T(t) := \exp\left(-\int_0^t \sigma(\mathbf{r}(s))ds\right) \tag{3}$$

where $T(t)$ is the accumulated transmittance along the ray $\mathbf{r}$. While Equation (2) provides the theoretically grounded manner to calculate the expected pixel color, the integration in Equation (2) is often intractable when $\mathbf{c}$ and σ are modeled as the MLP. Hence, numerical estimation [46] of the expected color is widely employed. The integration in Equation (2) is approximated using a set of N discrete points $\{p_i\}_{i=1}^N$ sampled along the ray $\mathbf{r}$, i.e., $p_i := \mathbf{o} + t_i\mathbf{d}$ for some $t_i \in (0, \infty)$. Then, the estimated color can be computed as follows,

$$\hat{C}(\mathbf{r}) = \sum_{i=1}^{N} w_i\mathbf{c}_i,\ w_i = T_i\alpha_i,\ T_i = \prod_{j=1}^{i-1}(1-\alpha_j), \tag{4}$$

where $\mathbf{c}_i$ and σ_i are a color and density at p_i, i.e., $(\mathbf{c}_i, \sigma_i) = f_\theta(p_i, \mathbf{d})$; w_i is a ray weight at p_i; T_i represents the accumulated transmittance; $\alpha_i = 1 - \exp(-\sigma_i\delta_i)$ represents the ray termination probability at p_i; and $\delta_i = t_{i+1} - t_i$ is the distance between adjacent samples. $\hat{C}(\mathbf{r})$ is computed as the weighted sum of color samples $\mathbf{c}_i$ with ray weight w_i.

Despite NeRF being highly effective in 3D scene representation, its dependency on volume density fields for novel view synthesis poses significant challenges for extracting high-quality surfaces. To mitigate this limitation, an existing approach combines neural implicit representation with signed distance functions (SDFs), leading to the development of neural SDF representation. In this framework, the SDF value $f(x)$ at a given 3D point x is modeled by the neural network. In [30], the employment of SDF values introduces a distinct formulation of the opacity term, α_i, different from the approach utilized in NeRF. Specifically, in [30], α_i is defined as

$$\alpha_i = \max\left(\frac{\Phi_s(f(p_i)) - \Phi_s(f(p_{i+1}))}{\Phi_s(f(p_i))}, 0\right), \tag{5}$$

where $f(p_i)$ represents the SDF value at 3D point p_i, and $\Phi_s(x) = (1 + e^{-sx})^{-1}$ denotes the sigmoid function. The parameter s is either automatically learned during the training process or manually set.

Due to the neural SDF representation, NeuS demonstrates strong capabilities in reconstructing the surfaces of novel objects. However, its application in real-time robotic systems is constrained by the time-consuming nature of the learning process and the significant computational costs involved. To address this limitation, we adopt Voxurf [35], a model that leverages voxel-based representations for accelerating the surface reconstruction process. Voxurf enhances the training speed by combining an explicit voxel grid with a neural SDF representation. Specifically, Voxurf directly learns the SDF values of the geometry through an SDF voxel grid $V^{(\mathrm{SDF})} \in \mathbb{R}^{1 \times N_x \times N_y \times N_z}$ and predicts the color using a feature voxel grid

$V^{(\text{feat})} \in \mathbb{R}^{C \times N_x \times N_y \times N_z}$ with a lightweight MLP. For a specific $(\mathbf{x}, \mathbf{d})$, the SDF value d' and the color $\mathbf{c}$ can be obtained using trilinear interpolation as follows:

$$d' = \text{interp}(\mathbf{x}, \mathcal{G}(V^{(\text{SDF})}, k_g, \sigma_g)), \tag{6}$$

$$\mathbf{c} = \text{MLP}(\text{interp}(\mathbf{x}, V^{(\text{feat})}), \mathbf{x}, \mathbf{d}, n). \tag{7}$$

For d', $\mathcal{G}$ represents a 3D convolution on the voxel grid $V^{(\text{SDF})}$ using a Gaussian kernel with size k_g and standard deviation σ_g to smooth the voxel SDF values rather than using the raw data. For $\mathbf{c}$, the normal vector n is derived from the $V^{(\text{SDF})}$. Lastly, d' is substituted into Equation (5) to calculate α_i and then volume rendering is performed as in Equation (4) to obtain the final pixel color.

4. Proposed Method

We propose the See-Then-Grasp (STG) method for active 3D modeling using a single robotic manipulator system. This method allows the robot to interact with the object, efficiently exploring and addressing uncertain regions for comprehensive 3D reconstruction. The pipeline consists of two main stages. The first stage is an accessible view collection. In this stage, the robot's hand-eye camera captures initial images from reachable viewpoints, building a preliminary understanding of the object's geometry. The second stage involves grasp-assisted full coverage. In the second stage, the robot grasps the object and uses a fixed external camera to capture additional images from previously inaccessible angles, ensuring complete 3D coverage. The entire pipeline is summarized in Figure 2.

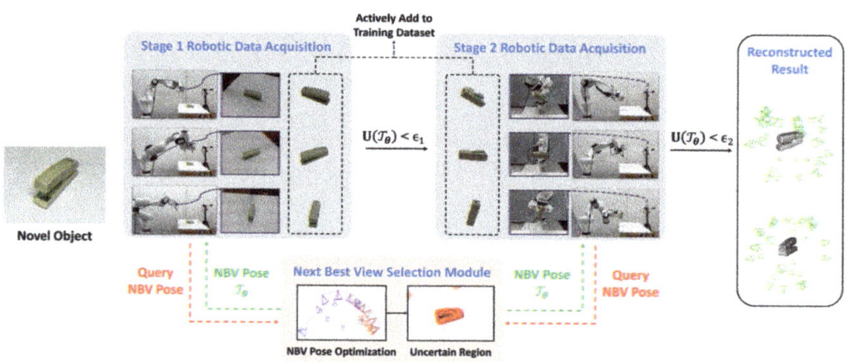

Figure 2. Overall pipeline. We propose the See-Then-Grasp method for learning the full 3D geometry of an object in a two-stage reconstruction process. In each stage, the robot actively acquires data by receiving the optimal next view pose $\mathcal{T}_{\Theta}$ from the next best view selection module, as represented by the red and green dashed lines, moving to that position, and capturing images. The captured images undergo post-processing and are actively added to the training dataset as represented by the black dashed lines, ultimately enabling the model to learn the full 3D geometry of the object.

4.1. Object Uncertainty Measurement

Unlike other uncertainty measurement methods [36,40,42] that require model modifications or extra parameter learning, we measure object geometry uncertainty using the entropy of the ray weight distribution. This entropy-based approach is efficient and fast, with no need for model changes.

$$H(\mathbf{r}) = -\sum_{i=1}^{N} [w_i \log(w_i) + (1 - w_i) \log(1 - w_i)], \tag{8}$$

where w_i is the ray weight defined in (4). Noisy ray weight distributions indicate inaccurate geometry, leading to high entropy. As geometry accuracy improves, the distribution converges to a single peak, reducing the entropy. Finally, the 2D pixel-wise uncertainty map $\mathbf{U}$, corresponding to the single camera pose $\mathcal{T}_{\text{cam}} \in \text{SE}(3)$, is calculated as the mean value of the entropy of all the rays:

$$\mathbf{U}(\mathcal{T}_{\text{cam}}) = \frac{1}{|\mathcal{R}_{\text{cam}}|} \sum_{\mathbf{r} \in \mathcal{R}_{\text{cam}}} H(\mathbf{r}), \tag{9}$$

where $\mathcal{R}_{\text{cam}}$ represents the batch of rays originating from single camera pose $\mathcal{T}_{cam}$.

While entropy-based uncertainty effectively represents object geometry uncertainty, applying it directly to the original NeRF model can cause overfitting, missing uncertainty in unseen views. To address this, we introduced an additional uncertain alpha voxel grid $V^{(unc,\alpha)}$, inspired by [44]. Initialized to maximum uncertainty, this grid is updated during training with α values from $V^{(\text{SDF})}$, maintaining high uncertainty in unseen viewpoints and enabling the NBV agent to capture diverse views. As illustrated in Figure 3, our uncertainty measurement can capture both seen-view and unseen-view uncertainties.

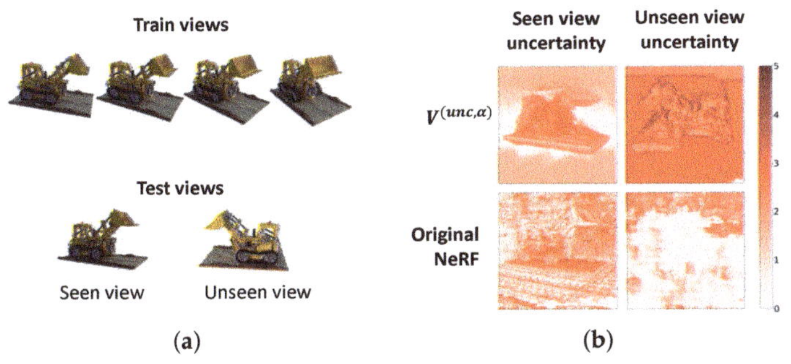

Figure 3. Comparison of uncertainty in the original NeRF and uncertain alpha voxel grid for the unseen view. We divide the views into (**a**) a seen view, which is similar to the train view, and an unseen view, which represents an out-of-distribution view, to investigate the overfitting issue of entropy-based uncertainty. As illustrated in (**b**), the original NeRF model fails to accurately capture the uncertainty in the unseen view due to the overfitting issue when trained using only one side of an object (**bottom right**). However, by applying $V^{(\text{unc},\alpha)}$ to a voxel-based model, it can effectively capture the uncertainty in the unseen view (**top right**).

4.2. Next Best View Pose Optimization

In robotic active reconstruction, quickly identifying the most uncertain view for NBV pose selection is crucial. The existing methods rely on fixed candidate views, requiring time-consuming rendering from all predefined viewpoints.

We propose a gradient-based NBV pose optimization method that reduces the rendering time by selecting NBV poses with fewer renderings. By using gradient information to pinpoint where uncertainty increases most rapidly, our approach achieves an efficient NBV with minimal renderings.

To generate an object-centric differentiable pose, we define the transformation matrix $\mathcal{T}_{\mathbf{\Theta}}$ using spherical coordinates (ϕ, θ, r) where $\mathbf{\Theta} := (\phi, \theta)$ indicates the learnable parameters. First, we define the translation vector $\mathbf{t}$ by converting the spherical coordinates to Cartesian coordinates:

$$\mathbf{t}_{\mathbf{\Theta}} = [x, y, z]^{\intercal} = [r \sin\theta \cos\phi, r \sin\theta \sin\phi, r \cos\theta]^{\intercal}. \tag{10}$$

Next, we define the rotation axes to ensure an object-centric view:

- The z-axis of the camera frame denoted as $\vec{w}$ is oriented toward the object origin:

$$\vec{w} = -\frac{(x, y, z)}{\sqrt{x^2 + y^2 + z^2}}. \tag{11}$$

- The x-axis of the camera frame denoted as $\vec{u}$ is aligned parallel to the xy-plane of the global object frame:

$$\vec{u} = \frac{(-y, x, 0)}{\sqrt{x^2 + y^2}}. \tag{12}$$

- The y-axis of the camera frame denoted as $\vec{v}$ is derived as the cross product of the rotation z-axis and the rotation x-axis:

$$\vec{v} = \vec{w} \times \vec{u}. \tag{13}$$

The rotation matrix $\mathbf{R}$ is formed by these orthogonal vectors, $\vec{u}$, $\vec{v}$, and $\vec{w}$:

$$\mathbf{R}_{\Theta} = \begin{pmatrix} \vec{u} & \vec{v} & \vec{w} \end{pmatrix}^{\mathsf{T}}. \tag{14}$$

Finally, the 4×4 transformation matrix $\mathcal{T}_{\Theta}$ is defined as follows:

$$\mathcal{T}_{\Theta} = \begin{pmatrix} \mathbf{R}_{\Theta} & \mathbf{t}_{\Theta} \\ 0 & 1 \end{pmatrix}. \tag{15}$$

By this construction, the transformation matrix $\mathcal{T}_{\Theta}$ is differentiable since all the pose construction steps are tractable.

To ensure diverse exploration during the NBV pose optimization process and avoid overfitting to the existing training set poses, we introduce a simple regularizer loss $\mathcal{L}_{\text{reg}}$ as follows:

$$\mathcal{L}_{\text{reg}} = \frac{1}{N_t} \sum_{i=1}^{N_t} \text{err}(\mathcal{T}_{\Theta}, \mathcal{T}_i), \tag{16}$$

where N_t is the number of poses in the existing training set, and $\text{err}(\cdot, \cdot)$ refers to the absolute value of the pose error, including both the euclidean distance between the translations and the rotational error between the orientations of the two poses.

The final NBV loss $\mathcal{L}_{\text{NBV}}$ is calculated as the negative sum of the uncertainty map as in Equation (9) observed from $\mathcal{T}_{\Theta}$ and $\mathcal{L}_{\text{reg}}$:

$$\mathcal{L}_{\text{NBV}} = -\lambda_1 \mathbf{U}(\mathcal{T}_{\Theta}) - \lambda_2 \mathcal{L}_{\text{reg}}, \tag{17}$$

where λ_1 and λ_2 are scaling parameters. By optimizing $\mathcal{L}_{\text{NBV}}$ with a gradient descent optimizer such as Adam [47], $\mathcal{T}_{\Theta}$ gradually moves toward viewpoints with increasing uncertainty within the current scene. As shown in Figure 4, our proposed NBV pose optimization enables direct optimization for uncertain viewpoints, allowing for 60% faster exploration compared to the fixed view candidate methodology.

After performing the optimization iterations, $\mathcal{T}_{\Theta}$ corresponds to the camera pose with the highest uncertainty. In other words, by moving the manipulator to satisfy this camera pose with respect to the novel object and actively updating the dataset, the 3D model can be reconstructed more quickly. A robotic pipeline capable of performing full 3D reconstruction of the object will be introduced in Section 4.3.

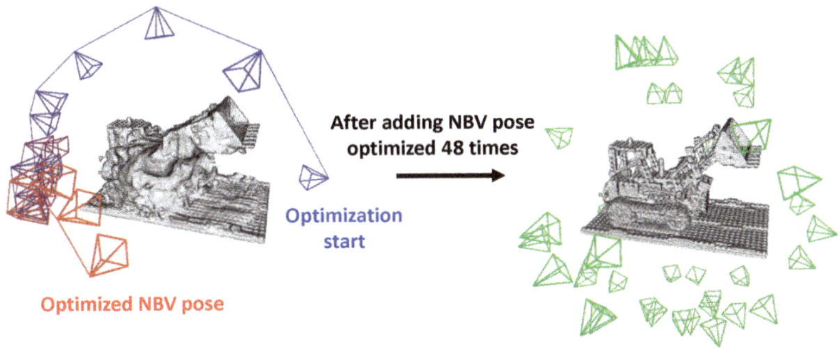

Figure 4. Visualization of our proposed gradient-based NBV pose optimization. Starting from the blue-colored pose added in the previous optimization step, we obtain a camera trajectory that moves to the uncertain red-colored pose while optimizing $\mathcal{L}_{\text{NBV}}$ (**left side**). Repeating this optimization step 50 times and adding it to the training set yields refined 3D reconstruction results (**right side**).

4.3. See-Then-Grasp: Two-Stage Robotic Full 3D Reconstruction Pipeline

Our proposed NBV pose optimization enables access to all viewpoints within a spherical view space for complete 3D reconstruction of an object. Previous studies, which used a fixed-base single manipulator system and a static object, often failed to achieve full coverage, especially for hidden surfaces. This system can only observe specific parts of the object, limiting the view space and preventing full 3D reconstruction of hidden surfaces like the bottom or back side. To overcome this limitation, we propose a novel two-stage reconstruction method utilizing a single robot manipulator and two cameras that directly interact with the object. The summarized method is represented in Algorithm 1.

Algorithm 1 See-Then-Grasp: Two-stage Robotic Active 3D Reconstruction

Require: Voxurf model, Initial training set $\{\mathcal{I}_i, \mathcal{T}_i\}_{i=1}^{1}$
Initialize Voxurf model and start training with $\{\mathcal{I}_i, \mathcal{T}_i\}_{i=1}^{1}$
Stage 1: Accessible View Collection
  Constrain NBV view space θ, ϕ
  while $\mathbf{U}(\mathcal{T}_\Theta) > \epsilon_1$ **do**
    if Training iteration mod $t_{\text{NBV}} = 0$ **then**
      Optimize $\mathcal{L}_{\text{NBV}}$ to find NBV pose $\mathcal{T}_\Theta$
      Move robot hand-eye camera to $\mathcal{T}_\Theta$
      Capture image $\mathcal{I}_t$ and apply mask $\mathcal{M}$
      Update training set $\{\mathcal{I}_i, \mathcal{T}_i\}_{i=1}^{t+1} \leftarrow \mathcal{I}_t, \mathcal{T}_\Theta$ and retrain model
    end if
    Optimize $\mathcal{L}_{\text{recon}}$
  end while
  Generate $\mathcal{T}_{\text{grasp\_eef}}^{\text{base}}$ and grasp the object
Stage 2: Grasp-Assisted Full Coverage
  Release the constraints θ, ϕ
  while $\mathbf{U}(\mathcal{T}_\Theta) > \epsilon_2$ **do**
    if Training iteration mod $t_{\text{NBV}} = 0$ **then**
      Optimize $\mathcal{L}_{\text{NBV}}$ to find NBV pose $\mathcal{T}_\Theta$
      Move robot end-effector to $\mathcal{T}_{\text{move}}$
      Capture image $\mathcal{I}_t$ and apply mask $\mathcal{M}$
      Update training set $\{\mathcal{I}_i, \mathcal{T}_i\}_{i=1}^{t+1+\text{stage 1}} \leftarrow \mathcal{I}_t, \mathcal{T}_\Theta$ and retrain model
    end if
    Optimize $\mathcal{L}_{\text{recon}}$
  end while

4.3.1. Stage 1 (Accessible View Collection)

Stage 1 involves utilizing the robot's hand-eye camera to perform the 3D reconstruction of objects within the robot's reachable view space. Specifically, the view space is constrained within the angular limits of $\theta_{\text{lim.1}} \leq \theta \leq \theta_{\text{lim.2}}$ and $\phi_{\text{lim.1}} \leq \phi \leq \phi_{\text{lim.2}}$, where $\theta_{\text{lim.1}}$, $\theta_{\text{lim.2}}$, $\phi_{\text{lim.1}}$, and $\phi_{\text{lim.2}}$ represent the maximum range the hand-eye camera can achieve. The primary objective of stage 1 is to preserve the areas that will be occluded by the gripper when the view space is expanded in stage 2.

At the beginning of stage 1, the Voxurf model starts training using the initial training dataset $\{\mathcal{I}_i, \mathcal{T}_i\}_{i=1}^{1}$, where $\mathcal{I}_i$ and $\mathcal{T}_i$ are augmented by selecting any view within the constrained view space. When the overall training iteration reaches every active learning iteration t_{NBV}, the NBV pose optimization begins. As a result of optimizing $\mathcal{L}_{\text{NBV}}$, the NBV pose $\mathcal{T}_{\mathbf{\Theta}}$ with the highest uncertainty is obtained. The robot hand-eye camera is then maneuvered to align with the object center frame according to the pose $\mathcal{T}_{\mathbf{\Theta}}$. Afterward, a new image is captured, and a segmentation mask is applied to acquire the NBV image data $\mathcal{I}_t$. The new image and pose pair are added to the current training dataset, updating it to $\{\mathcal{I}_i, \mathcal{T}_i\}_{i=1}^{t+1} \leftarrow \mathcal{I}_t, \mathcal{T}_{\mathbf{\Theta}}$. Because the training dataset is updated through active learning, the entire training ray batch $\mathcal{R}_{\text{train}}$ is also augmented with rays from the updated pose set.

Finally, the reconstruction loss is calculated using the updated training ray batch $\mathcal{R}_{\text{train}}$ as follows:

$$\mathcal{L}_{\text{recon}} = \frac{1}{|\mathcal{R}_{\text{train}}|} \sum_{r \in \mathcal{R}_{\text{train}}} (||\tilde{C}(\mathbf{r}), \hat{C}(\mathbf{r})||_2^2), \tag{18}$$

where the expected pixel color $\hat{C}(\mathbf{r})$ is computed as in Equation (4), while $\tilde{C}(\mathbf{r})$ represents the ground truth pixel color for ray $\mathbf{r}$ and can be obtained from the training image dataset $\mathcal{I}_i$. In our active learning setting, data are gradually added for the uncertain regions of the object, resulting in the progressive improvement in the object's shape as the reconstruction loss $\mathcal{L}_{\text{recon}}$ is optimized. The active learning process continues until the object uncertainty $\mathbf{U}(\mathcal{T}_{\mathbf{\Theta}})$ falls below a predefined threshold ϵ_1, at which point the system transitions to stage 2.

4.3.2. Stage 2 (Grasp-Assisted Full Coverage)

In stage 2, the manipulator grasps the object and utilizes another fixed external camera to cover all areas of the object by showing the uncertain regions. Specifically, to obtain data regarding the NBV pose $\mathcal{T}_{\mathbf{\Theta}}$ that allows for capturing the complete view of the object, the robot grasps the object and moves its end effector, satisfying $\mathcal{T}_{\mathbf{\Theta}}$ relative to the external camera. The robot first employs a grasp pose generator such as Contact-GraspNet [48] to generate the grasp pose $\mathcal{T}_{\text{grasp\_eef}}^{\text{base}}$ for grasping the object. Subsequently, we release the view space constraints from stage 1, expanding the view space around the object to a full spherical range. Then, the robot end-effector pose $\mathcal{T}_{\text{move}}^{\text{base}}$ is generated following the transformation order.

$$\mathcal{T}_{\text{move}}^{\text{base}} = \mathcal{T}_{\text{cam2}}^{\text{base}} (\mathcal{T}_{\mathbf{\Theta}})^{-1} (\mathcal{T}_{\text{obj}}^{\text{base}})^{-1} \mathcal{T}_{\text{grasp\_eef}}^{\text{base}}, \tag{19}$$

where $\mathcal{T}_{\text{cam2}}^{\text{base}}$ and $\mathcal{T}_{\text{obj}}^{\text{base}}$ represent the pose of the external camera and the object canonical pose placed on the table, both relative to the robot's base frame. Note that $(\mathcal{T}_{\text{obj}}^{\text{base}})^{-1} \mathcal{T}_{\text{grasp\_eef}}^{\text{base}}$ represents the transformation between the object pose and the end-effector frame after the object has been grasped and attached. By planning the path of the end effector according to $\mathcal{T}_{\text{move}}^{\text{base}}$ while the robot grasps the object, the object pose as seen from the external camera can satisfy $\mathcal{T}_{\mathbf{\Theta}}$.

When stage 2 begins, the Voxurf model initiates the active learning process, incorporating the training dataset obtained from stage 1. By optimizing $\mathcal{L}_{\text{NBV}}$, the full-coverage

NBV pose $\mathcal{T}_{\mathbf{\Theta}}$ is determined, and the robot's end effector is moved to $\mathcal{T}_{\text{move}}^{\text{base}}$ to align with $\mathcal{T}_{\mathbf{\Theta}}$ as seen from the external camera. Subsequently, an image is captured with the external camera and a segmentation mask is applied, adding the new image and pose pair to the current training dataset: $\{\mathcal{I}_i, \mathcal{T}_i\}_{i=1}^{t+1+\text{stage 1}} \leftarrow \mathcal{I}_t, \mathcal{T}_{\mathbf{\Theta}}$. Similar to stage 1, the training process continues by optimizing the reconstruction loss $\mathcal{L}_{\text{recon}}$ in Equation (18). The process is concluded when the object uncertainty $\mathbf{U}(\mathcal{T}_{\mathbf{\Theta}})$ drops below the predefined threshold ϵ_2.

Due to the use of two cameras, the backgrounds of the views from each camera differ, making it challenging to satisfy the view consistency of the object. To ensure the view consistency for each camera, we applied HQ-SAM [49] to generate object segmentation masks from the captured images. Furthermore, to retain the geometric information obtained in stage 1 during stage 2, we prevent gradients from flowing in areas outside the mask, thus preserving the information in regions occluded by the gripper. Specifically, a binary segmentation mask $\mathcal{M}$, which has the same size as the raw image $\mathcal{I}_{\text{raw}}$, is defined as follows:

$$\mathcal{M}(\mathcal{I}_{\text{raw}}) = \begin{cases} 1 & \text{if the pixel in } \mathcal{I}_{\text{raw}} \in \text{object}, \\ 0 & \text{if the pixel in } \mathcal{I}_{\text{raw}} \in \text{background}. \end{cases} \quad (20)$$

At each stage, after generating a mask that excludes the background areas, including non-object regions such as the robot gripper and the surrounding environment, the mask is applied to the original image to create the image used for training, $\mathcal{I}$, as follows:

$$\mathcal{I} = \mathcal{M}(\mathcal{I}_{\text{raw}}) \cdot \mathcal{I}_{\text{raw}}. \quad (21)$$

By using the masked images for training, the object images generated at each stage maintain view consistency, allowing the Voxurf model to focus on learning the objects. As a result, the data acquired from each stage enable the complete 3D reconstruction of the object.

5. Experiment

We conduct experiments on synthetic and real-world objects to evaluate our proposed method for active 3D reconstruction through NBV pose optimization. In Section 5.1, we validate the proposed uncertainty measurement method and NBV pose optimization on synthetic objects. In Section 5.2, we describe the experimental results of applying the proposed STG method to a real robot for novel real-world objects.

5.1. Experiment on Synthetic Objects

5.1.1. Implementation Details

To implement the proposed NBV pose optimization, we set the pose parameters ϕ and θ as learnable parameters using PyTorch (version 2.0) [50] while fixing the radius r at 4.0. We use the Adam [47] optimizer with a learning rate of 1.0 to optimize the NBV loss function $\mathcal{L}_{\text{NBV}}$. Additionally, the scaling parameters λ_1 and λ_2 are set to 1.0 and 0.5, respectively. For each optimization step, we perform 20 optimization iterations every 2000 training iterations to obtain the NBV pose. The experiments utilize the code provided by the authors of Voxurf [35]. Since Voxurf consists of a coarse-to-fine stage training procedure, the data obtained from NBV optimization in the coarse stage are used for training in the fine stage to achieve the final 3D reconstruction results. All the synthetic dataset experiments are conducted on a system equipped with an Intel Xeon Gold 6226R CPU at 2900 MHz, 125 GB of RAM, and an NVIDIA GeForce RTX A5000 GPU. The CPU is manufactured by Intel Corporation (Santa Clara, CA, USA), and the GPU is manufactured by NVIDIA Corporation (Santa Clara, CA, USA).

5.1.2. Dataset

We select four objects—Lego, Ficus, Chair, and Mic—from the NeRF-Synthetic [11] dataset. For each NBV pose optimization step, we use Blender [51] to render the image of the optimized pose to generate the training image data. Additionally, we generate 150 images and corresponding pose data that cover the entire view of each object, including the bottom side, for use as the test dataset.

5.1.3. Metrics

We evaluate the results of the synthetic experiments in terms of rendering quality and geometric quality. For rendering quality, we use the peak signal-to-noise ratio (PSNR) and the learned perceptual image patch similarity (LPIPS) metric. The PSNR measures the fidelity of the rendered 2D images by calculating the pixel-wise difference between the images rendered from the trained model and the ground truth images, with higher PSNR values indicating better image quality. LPIPS, on the other hand, evaluates perceptual similarity using a deep learning-based model, where lower LPIPS values indicate greater visual similarity to the ground truth images.

For geometric quality, we use the F-score and Chamfer distance metrics to assess the difference between the reconstructed 3D mesh and the ground truth mesh. The F-score is calculated by sampling points from each mesh and setting a threshold of 1 cm to evaluate completeness and accuracy, where higher F-score values indicate better alignment between the reconstructed and ground truth meshes. The Chamfer distance measures the average distance between points on the reconstructed mesh and their closest counterparts on the ground truth mesh, with smaller values indicating a closer match and thus higher geometric fidelity.

5.1.4. Results

The quantitative results of our experiments are presented in Table 1, while the qualitative results of our baseline are shown in Figure 5. According to Table 1, our baseline outperforms other baselines in terms of rendering and geometric quality when 24 NBV images are added. This performance difference is particularly significant for geometrically complex datasets, such as Lego and Ficus. Even with the addition of 48 NBV images, our baseline demonstrates superior results, although the other baselines tend to converge to a similar level.

Figure 5. Qualitative 3D reconstruction results of our baseline. We present the 3D reconstruction of four synthetic objects, improved by adding 48 NBV images to our baseline. The ground truth images are on the left, with the 3D reconstruction meshes on the right, showcasing our method's ability to accurately capture fine details.

To further evaluate the NBV selection performance of each baseline, we present the qualitative comparison results in Figure 6 and the F-score progression for the Ficus dataset in Figure 7. These results indicate that our baseline reaches the oracle line more rapidly than other methods and effectively captures both the geometric details and the overall shape.

Table 1. Quantitative results of synthetic object reconstruction. We quantitatively evaluated the performance of NBV selection using four synthetic objects. The experiments were conducted under two scenarios: using 24 NBV images and using 48 NBV images. Each experiment was repeated five times with different random seeds, and the average results are reported. Grad denotes our NBV pose optimization method, whereas Fixed refers to a fixed pose candidate method. Note that CD represents the Chamfer distance, with the results being scaled by a factor of 100. Higher values are better for PSNR and F-Score, while lower values are better for LPIPS and CD.

	Lego				Ficus				Chair				Mic			
Method	**PSNR ↑**	**LPIPS ↓**	**F-Score ↑**	**CD ↓**	**PSNR ↑**	**LPIPS ↓**	**F-Score ↑**	**CD ↓**	**PSNR ↑**	**LPIPS ↓**	**F-Score ↑**	**CD ↓**	**PSNR ↑**	**LPIPS ↓**	**F-Score ↑**	**CD ↓**
Quality after 24 NBV images added																
Grad + Entropy (Ours)	31.69	0.0508	0.4022	0.2613	26.96	0.0564	0.6403	0.1071	31.30	0.0319	0.3810	0.3345	30.26	0.0312	0.4010	0.3045
Grad + Ensemble [42]	30.48	0.0637	0.3908	0.2708	24.24	0.0780	0.6129	0.1479	31.46	0.0312	0.3891	0.3168	28.58	0.0319	0.4091	0.2868
Fixed [17,18] + Entropy	29.29	0.0682	0.3651	0.3665	25.41	0.0544	0.5308	0.1899	30.13	0.0394	0.3887	0.3214	29.11	0.0394	0.4098	0.2848
Fixed [17,18] + Ensemble [42]	30.26	0.0522	0.3819	0.6322	24.11	0.0798	0.6184	0.1381	31.85	0.0329	0.3898	0.3148	27.73	0.0389	0.4087	0.2914
FVS	28.63	0.0611	0.3509	0.4427	24.52	0.0539	0.5003	0.2215	29.97	0.0464	0.3688	0.4269	26.15	0.0464	0.3888	0.4169
Random	27.46	0.0789	0.2649	0.7026	23.26	0.0671	0.4834	0.3083	28.56	0.0641	0.3464	0.5938	26.30	0.0641	0.3664	0.5938
Quality after 48 NBV images added																
Grad + Entropy (Ours)	34.73	0.0319	0.4886	0.1408	34.49	0.0204	0.8482	0.0504	35.86	0.0275	0.4168	0.2844	34.15	0.0230	0.7241	0.0785
Grad + Ensemble [42]	34.04	0.0321	0.4711	0.1425	34.13	0.0244	0.8149	0.1048	36.10	0.0248	0.4216	0.2805	34.79	0.0202	0.7211	0.0736
Fixed [17,18] + Entropy	34.16	0.0324	0.4295	0.2857	34.51	0.0203	0.8405	0.0514	35.60	0.0294	0.4144	0.2869	33.73	0.0229	0.7232	0.0787
Fixed [17,18] + Ensemble [42]	34.10	0.0363	0.4603	0.1616	30.08	0.0715	0.8198	0.0945	36.04	0.0253	0.4128	0.2833	34.41	0.0212	0.7181	0.0791
FVS	32.79	0.0384	0.4413	0.2134	32.85	0.0289	0.8356	0.1101	35.27	0.0271	0.4135	0.2845	34.02	0.0234	0.7178	0.0814
Random	33.41	0.0390	0.4118	0.2772	33.62	0.0237	0.8301	0.1148	35.59	0.0263	0.4121	0.2883	33.89	0.0241	0.7072	0.0836
All Images	-	-	0.4992	0.095	-	-	0.8560	0.0476	-	-	0.4244	0.2615	-	-	0.7279	0.0715

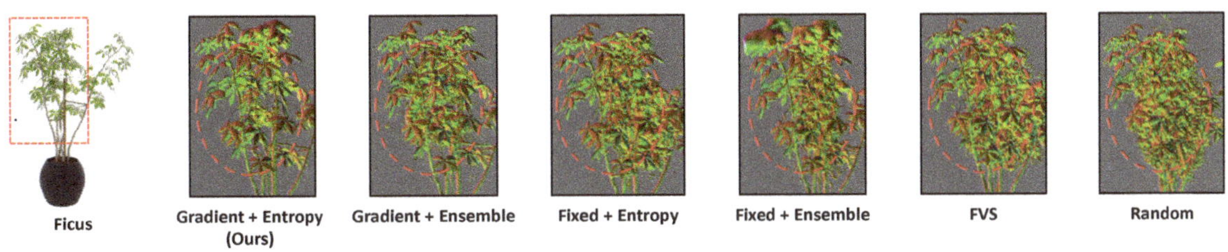

Figure 6. Qualitative comparison of the baselines on the Ficus dataset. We present the surface normal plots of the 3D models using 24 NBV images from the Ficus dataset. The red dashed line shows that our baseline effectively captures geometric details, while others produce noisy geometries.

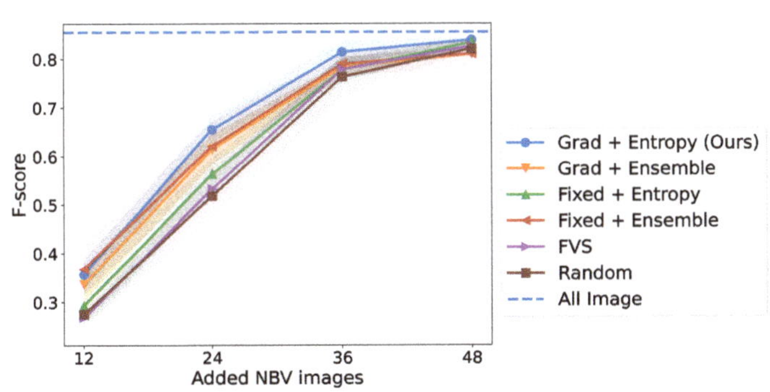

Figure 7. F-score improvement plot for the baselines on the Ficus dataset. We analyze the F-score improvement on the Ficus dataset using 12, 24, 36, and 48 NBV images, with the results averaged over five random seeds. The mean-variance plots show that our baseline quickly identifies regions that significantly boost the F-score, especially in the early stages of NBV selection.

Finally, the time cost measurements for each baseline are illustrated in Figure 8. Our NBV pose optimization efficiently identifies uncertain regions without exploring all view candidates through a single NBV pose optimization. As a result, our method solves the NBV selection problem 60% faster compared to the fixed view candidate method.

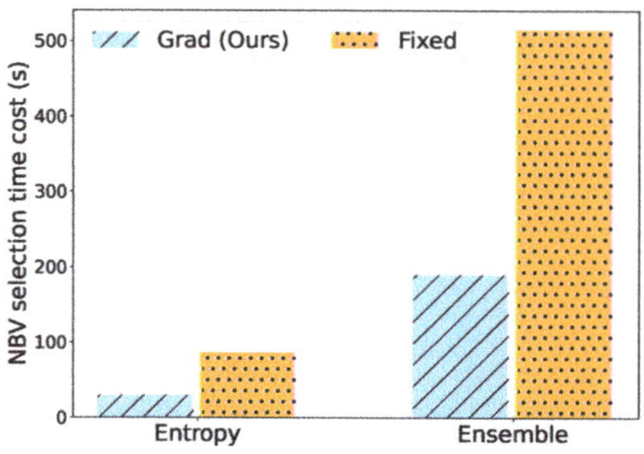

Figure 8. Comparison of NBV selection time costs. We measured the time for each NBV selection across different uncertainty techniques and methods. The sky blue textured bar represents the time taken when performing our NBV pose optimization, while the orange textured bar represents the time taken when selecting the NBV pose using fixed view candidates. Our method consistently incurs lower time costs compared to the fixed candidate method.

5.2. Experiment on Real-World Objects

5.2.1. Implementation Details

We apply the proposed two-stage reconstruction method to a real robot to achieve full reconstruction of real-world objects. The overall NBV pose optimization approach follows the same settings as defined in Section 5.1. In stage 1, to generate $\mathcal{T}_{\Theta}$ within the robot's reachable range, we set $\theta_{\text{lim.1}}$ and $\theta_{\text{lim.2}}$ to $\pi/5$ and $\pi/3$, respectively, and $\phi_{\text{lim.1}}$ and $\phi_{\text{lim.2}}$ to $7\pi/6$ and $11\pi/6$, respectively. An HQ-SAM [49] model with pretrained ViT-L [52] is employed to create segmentation masks for the objects at each stage. The stop criteria for each stage, ϵ_1 and ϵ_2, are set to 0.5 and 1.0, respectively.

We utilize the Franka Panda manipulator (Franka Emika GmbH, München, Germany) as the robotic hardware platform and employ two RealSense D435i cameras (Intel Corporation, Santa Clara, CA, USA), with one configured as a robot hand-eye camera and the other as an external fixed camera. For collision-free motion planning of the robot, the RRT-Connect [53] algorithm is employed. All the real-world experiments are conducted on a system equipped with a 12th Gen Intel Core i7 CPU at 3600 MHz, 32 GB of RAM, and an NVIDIA GeForce RTX 3080 GPU.

5.2.2. Datasets

We select four real-world objects: Mouse, Toy car, Stapler, and Toy screw. During stages 1 and 2, images of the objects are captured using cameras set to a 640 × 480 resolution. Conducting experiments on novel real-world objects without ground truth 3D models presents significant challenges in obtaining accurate object poses. Therefore, we predefine $\mathcal{T}_{\text{move}}$ with the help of an AprilTag [54] cube, ensuring that during training the obtained $\mathcal{T}_{\text{move}}$ aligns with the nearest predefined poses. We pregenerate 300 $\mathcal{T}_{\text{move}}$ to cover all the viewpoints of the object, ensuring that the maximum error between the poses generated during training and the pregenerated poses remains below 0.1.

5.2.3. Results

We present the qualitative results using our proposed STG method for each real object in Figure 9. Our method allows the robot to interact directly with the object, capturing all viewpoints, including the bottom side, which are not visible from a robot hand-eye camera. As illustrated in Figure 9, our method effectively reconstructs the back and bottom of objects, even in a limited view space with a fixed-based manipulator setup.

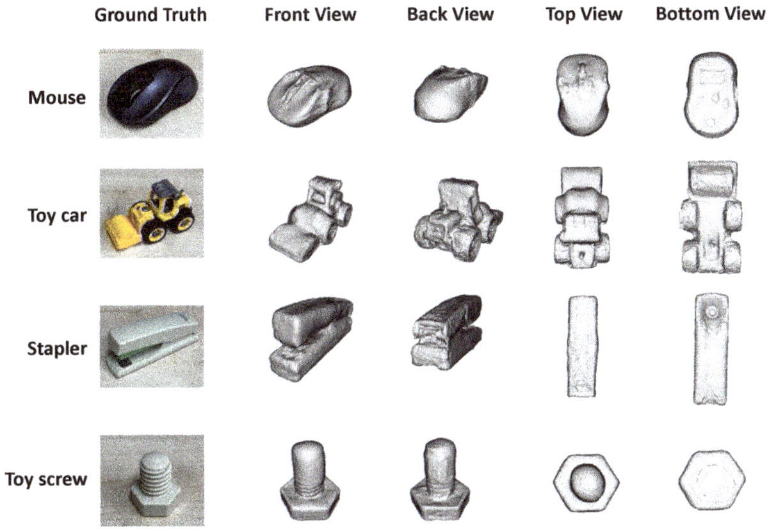

Figure 9. Qualitative results of 3D models for four real-world objects. We present 3D reconstructions of novel objects using our STG method, capturing all viewpoints, including the bottom side.

As demonstrated in Section 5.1, our gradient-based NBV pose optimization significantly reduces the number of rendering iterations required to identify uncertain regions compared to the fixed view candidate method. Under the same experimental conditions, the real-world experiments showed consistent results, achieving a 60% improvement in efficiency for identifying uncertain regions. This efficiency translates into a total average time of approximately 15 min for NBV data collection and full 3D reconstruction, including the robot's operation time. These results demonstrate that our proposed STG method enables robots to rapidly acquire the necessary training data for the comprehensive 3D reconstruction of objects, maintaining both speed and reconstruction quality across synthetic and real-world scenarios.

6. Ablation Study

6.1. Performance Comparison Based on NBV Pose Optimizer Parameters

Since we use the Adam [47] optimizer for NBV pose optimization, we report the comparative results in Table 2, obtained by varying the learning rate and optimization steps of the optimizer. Our NBV pose optimization explores the entire sphere; so with a low learning rate, the exploration range is reduced, requiring many optimization steps to reach uncertain areas. Therefore, we increase the learning rate to 1.0 to quickly reach uncertain areas with fewer optimization steps. Intuitively, more optimization steps lead to a more accurate approach to uncertain areas, but this comes with the trade-off of increased time costs. Thus, we utilize the optimizer configuration that achieves acceptable results with only 20 optimization steps, enabling an efficient approach to uncertain regions with a relatively low time cost.

Table 2. Quantitative results of the NBV pose optimizer with adjusted parameters for the Ficus dataset. We report the quantitative results from varying the learning rate and optimization steps of the NBV pose optimizer on the Ficus dataset. Our configuration achieves acceptable results with a reduced time cost compared to other settings. Higher values are better for PSNR and F-Score, while lower values are better for LPIPS, CD and time costs.

Learning Rate/Optimization Step	PSNR ↑	LPIPS ↓	F-Score ↑	CD ↓	Time Cost (s) ↓
0.1/10	23.13	0.0771	0.5185	0.2498	13.41
0.1/20	23.19	0.0702	0.5548	0.1848	30.04
0.1/40	23.42	0.0689	0.5741	0.1754	60.85
1.0/10	25.49	0.0607	0.5713	0.1248	14.85
1.0/20 (Ours)	26.96	0.0564	0.6403	0.1071	29.69
1.0/40	27.89	0.0479	0.6387	0.1046	60.31

6.2. Effectiveness of Our Two-Stage Reconstruction Pipeline

To validate the effectiveness of our two-stage method, we provide the qualitative results of iterative 3D reconstructions at each stage in Figure 10. As shown in Figure 10, during stage 1, where 4 and 12 NBV images are added, the approximate geometry of the Toy car can be captured from the upper view. However, the geometry from the lower view, which is inaccessible to the view space, cannot be adequately captured. Subsequently, when 24 NBV images are added, the data from stage 2 begin to improve the geometry of the bottom view. When 36 NBV images are added, the geometric details from all views can be fully captured. Thus, our STG method effectively achieves a comprehensive 3D reconstruction of the object.

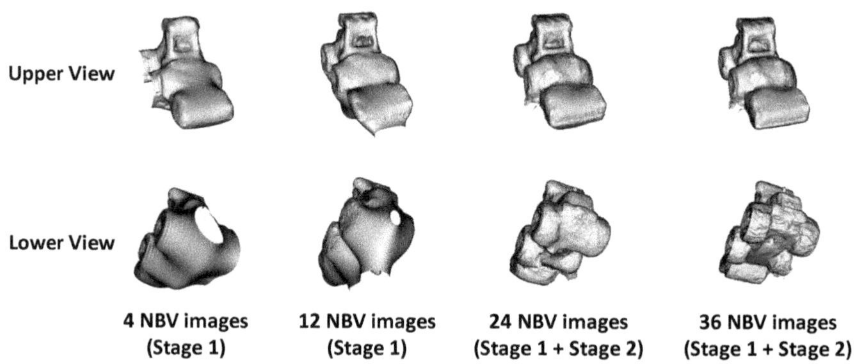

Figure 10. Iterative 3D reconstruction results for the Toy car dataset. We present the qualitative results of the iterative 3D reconstruction on the Toy car dataset, showing that our method progressively captures both upper and lower views.

7. Conclusions

This paper tackles the problem of active robotic 3D reconstruction for novel objects using a single robot manipulator system. We proposed a NBV pose optimization method that efficiently identifies uncertain regions of the object using a gradient-based approach. Our pose optimization method, which directly moves to uncertain regions, offers a notable improvement in time efficiency compared to the fixed candidate approach that evaluates all pose candidates. To address the limited viewpoint issue of a single robot manipulator system, we introduced the See-Then-Grasp (STG) method, a two-stage robotic full 3D reconstruction pipeline. Initially, the robot learns the approximate geometry from reachable viewpoints, then grasps the object and expands the viewpoint by showing the uncertain region to an external fixed camera. We conducted experiments on both synthetic and real-world objects, demonstrating the effectiveness of our approach.

Our results confirm that the proposed NBV pose optimization directs the robot to uncertain regions more quickly and accurately than other baseline methods, achieving a 60% improvement in computational efficiency. Furthermore, applying our method to real-world objects validates its capability to achieve full 3D reconstruction, including challenging occluded areas such as the bottom and back sides of objects. These findings underscore the potential of our method to significantly enhance the efficiency and quality of robotic 3D reconstruction in diverse environments.

While this work successfully addresses several challenges in active robotic 3D reconstruction, it is limited by its reliance on a single manipulator and the inherent constraints of fixed-base systems. Future work will focus on extending this approach to multi-robot systems and exploring real-time 6D pose estimation [55,56] to further improve reconstruction performance. By providing a robust framework for autonomous and complete object modeling, our method opens new opportunities for applications in manufacturing, logistics, and robotics research.

Author Contributions: Conceptualization, Y.H., J.K. and K.L.; methodology, Y.H. and K.L.; software, Y.H. and J.K.; validation, Y.H.; writing—original draft preparation, Y.H., J.K. and K.L.; writing—review and editing, Y.H., G.C., E.K. and K.L.; supervision, G.C., E.K. and K.L. All authors have read and agreed to the published version of the manuscript.

Funding: This work was supported by the National Research Foundation of Korea (NRF) grant funded by the Korea government (MSIT) (No. RS-2023-00211357) and the Chung-Ang University Graduate Research Scholarship in 2023.

Institutional Review Board Statement: Not applicable.

Informed Consent Statement: Informed consent was obtained from all the subjects involved in this study.

Data Availability Statement: The data supporting the findings of this study are available upon request from the corresponding author. Public access is restricted to protect privacy.

Conflicts of Interest: Author Geonho Cha was employed by the company NAVER Cloud. The remaining authors declare that the research was conducted in the absence of any commercial or financial relationships that could be construed as a potential conflict of interest.

References

1. Ota, K.; Jha, D.K.; Oiki, T.; Miura, M.; Nammoto, T.; Nikovski, D.; Mariyama, T. Trajectory optimization for unknown constrained systems using reinforcement learning. In Proceedings of the 2019 IEEE/RSJ International Conference on Intelligent Robots and Systems (IROS), Macau, China, 3–8 November 2019; pp. 3487–3494.
2. Luo, J.; Solowjow, E.; Wen, C.; Ojea, J.A.; Agogino, A.M.; Tamar, A.; Abbeel, P. Reinforcement learning on variable impedance controller for high-precision robotic assembly. In Proceedings of the 2019 International Conference on Robotics and Automation (ICRA), Montreal, QC, Canada, 20–24 May 2019; pp. 3080–3087.
3. Sun, Y.; Wang, W.; Chen, Y.; Jia, Y. Learn how to assist humans through human teaching and robot learning in human–robot collaborative assembly. *IEEE Trans. Syst. Man Cybern. Syst.* **2020**, *52*, 728–738. [CrossRef]
4. Liang, J.; Huang, W.; Xia, F.; Xu, P.; Hausman, K.; Ichter, B.; Florence, P.; Zeng, A. Code as policies: Language model programs for embodied control. In Proceedings of the 2023 IEEE International Conference on Robotics and Automation (ICRA), London, UK, 29 May–2 June 2023; pp. 9493–9500.

5. Zitkovich, B.; Yu, T.; Xu, S.; Xu, P.; Xiao, T.; Xia, F.; Wu, J.; Wohlhart, P.; Welker, S.; Wahid, A.; et al. Rt-2: Vision-language-action models transfer web knowledge to robotic control. In Proceedings of the 7th Conference on Robot Learning, PMLR, Atlanta, GA, USA, 6 November 2023; pp. 2165–2183.
6. Liu, Z.; Bahety, A.; Song, S. REFLECT: Summarizing Robot Experiences for Failure Explanation and Correction. In Proceedings of the 7th Conference on Robot Learning, PMLR, Atlanta, GA, USA, 6 November 2023; pp. 3468–3484.
7. Tachella, J.; Altmann, Y.; Mellado, N.; McCarthy, A.; Tobin, R.; Buller, G.S.; Tourneret, J.Y.; McLaughlin, S. Real-time 3D reconstruction from single-photon lidar data using plug-and-play point cloud denoisers. *Nat. Commun.* **2019**, *10*, 4984. [CrossRef] [PubMed]
8. Wu, Q.; Yang, H.; Wei, M.; Remil, O.; Wang, B.; Wang, J. Automatic 3D reconstruction of electrical substation scene from LiDAR point cloud. *ISPRS J. Photogramm. Remote Sens.* **2018**, *143*, 57–71. [CrossRef]
9. Wu, C.; Zeng, R.; Pan, J.; Wang, C.C.; Liu, Y.J. Plant phenotyping by deep-learning-based planner for multi-robots. *IEEE Robot. Autom. Lett.* **2019**, *4*, 3113–3120. [CrossRef]
10. Wu, S.; Sun, W.; Long, P.; Huang, H.; Cohen-Or, D.; Gong, M.; Deussen, O.; Chen, B. Quality-driven poisson-guided autoscanning. *ACM Trans. Graph. (TOG)* **2014**, *33*, 1–12. [CrossRef]
11. Mildenhall, B.; Srinivasan, P.P.; Tancik, M.; Barron, J.T.; Ramamoorthi, R.; Ng, R. Nerf: Representing scenes as neural radiance fields for view synthesis. *Commun. ACM* **2021**, *65*, 99–106. [CrossRef]
12. Adamkiewicz, M.; Chen, T.; Caccavale, A.; Gardner, R.; Culbertson, P.; Bohg, J.; Schwager, M. Vision-only robot navigation in a neural radiance world. *IEEE Robot. Autom. Lett.* **2022**, *7*, 4606–4613. [CrossRef]
13. Kwon, O.; Park, J.; Oh, S. Renderable neural radiance map for visual navigation. In Proceedings of the IEEE/CVF Conference on Computer Vision and Pattern Recognition, Vancouver, BC, Canada, 17–24 June 2023; pp. 9099–9108.
14. Zhu, Z.; Peng, S.; Larsson, V.; Xu, W.; Bao, H.; Cui, Z.; Oswald, M.R.; Pollefeys, M. Nice-slam: Neural implicit scalable encoding for slam. In Proceedings of the IEEE/CVF Conference on Computer Vision and Pattern Recognition, New Orleans, LA, USA, 18–24 June 2022; pp. 12786–12796.
15. Dai, Q.; Zhu, Y.; Geng, Y.; Ruan, C.; Zhang, J.; Wang, H. Graspnerf: Multiview-based 6-dof grasp detection for transparent and specular objects using generalizable nerf. In Proceedings of the 2023 IEEE International Conference on Robotics and Automation (ICRA), London, UK, 29 May–2 June 2023; pp. 1757–1763.
16. Blukis, V.; Lee, T.; Tremblay, J.; Wen, B.; Kweon, I.S.; Yoon, K.J.; Fox, D.; Birchfield, S. One-Shot Neural Fields for 3D Object Understanding. In Proceedings of the CVPR Workshop on Advances in NeRF for the Metaverse (XRNeRF), Vancouver, BC, Canada, 18 June 2023.
17. Lee, S.; Chen, L.; Wang, J.; Liniger, A.; Kumar, S.; Yu, F. Uncertainty guided policy for active robotic 3d reconstruction using neural radiance fields. *IEEE Robot. Autom. Lett.* **2022**, *7*, 12070–12077. [CrossRef]
18. Ran, Y.; Zeng, J.; He, S.; Chen, J.; Li, L.; Chen, Y.; Lee, G.; Ye, Q. Neurar: Neural uncertainty for autonomous 3d reconstruction with implicit neural representations. *IEEE Robot. Autom. Lett.* **2023**, *8*, 1125–1132. [CrossRef]
19. Yan, D.; Liu, J.; Quan, F.; Chen, H.; Fu, M. Active Implicit Object Reconstruction using Uncertainty-guided Next-Best-View Optimization. *IEEE Robot. Autom. Lett.* **2023**, *8*, 6395–6402. [CrossRef]
20. Zeng, J.; Li, Y.; Ran, Y.; Li, S.; Gao, F.; Li, L.; He, S.; Chen, J.; Ye, Q. Efficient view path planning for autonomous implicit reconstruction. In Proceedings of the 2023 IEEE International Conference on Robotics and Automation (ICRA), London, UK, 29 May–2 June 2023; pp. 4063–4069.
21. Monica, R.; Aleotti, J. A probabilistic next best view planner for depth cameras based on deep learning. *IEEE Robot. Autom. Lett.* **2021**, *6*, 3529–3536. [CrossRef]
22. Pan, S.; Jin, L.; Hu, H.; Popović, M.; Bennewitz, M. How many views are needed to reconstruct an unknown object using nerf? In Proceedings of the 2024 IEEE International Conference on Robotics and Automation (ICRA), Yokohama, Japan, 13–17 May 2024; pp. 12470–12476.
23. Hardouin, G.; Moras, J.; Morbidi, F.; Marzat, J.; Mouaddib, E.M. A multirobot system for 3-D surface reconstruction with centralized and distributed architectures. *IEEE Trans. Robot.* **2023**, *39*, 2623–2638. [CrossRef]
24. Lee, I.D.; Seo, J.H.; Kim, Y.M.; Choi, J.; Han, S.; Yoo, B. Automatic pose generation for robotic 3-D scanning of mechanical parts. *IEEE Trans. Robot.* **2020**, *36*, 1219–1238. [CrossRef]
25. Delmerico, J.; Isler, S.; Sabzevari, R.; Scaramuzza, D. A comparison of volumetric information gain metrics for active 3D object reconstruction. *Auton. Robot.* **2018**, *42*, 197–208. [CrossRef]
26. Newcombe, R.A.; Izadi, S.; Hilliges, O.; Molyneaux, D.; Kim, D.; Davison, A.J.; Kohi, P.; Shotton, J.; Hodges, S.; Fitzgibbon, A. Kinectfusion: Real-time dense surface mapping and tracking. In Proceedings of the 2011 10th IEEE International Symposium on Mixed and Augmented Reality, Basel, Switzerland, 26–29 October 2011; pp. 127–136.
27. Tychola, K.A.; Tsimperidis, I.; Papakostas, G.A. On 3D reconstruction using rgb-d cameras. *Digital* **2022**, *2*, 401–421. [CrossRef]

28. Cai, Z.; Han, J.; Liu, L.; Shao, L. RGB-D datasets using microsoft kinect or similar sensors: A survey. *Multimed. Tools Appl.* **2017**, *76*, 4313–4355. [CrossRef]
29. Barron, J.T.; Mildenhall, B.; Verbin, D.; Srinivasan, P.P.; Hedman, P. Mip-nerf 360: Unbounded anti-aliased neural radiance fields. In Proceedings of the IEEE/CVF Conference on Computer Vision and Pattern Recognition, New Orleans, LA, USA, 18–24 June 2022; pp. 5470–5479.
30. Wang, P.; Liu, L.; Liu, Y.; Theobalt, C.; Komura, T.; Wang, W. NeuS: Learning Neural Implicit Surfaces by Volume Rendering for Multi-view Reconstruction. *Adv. Neural Inf. Process. Syst.* **2021**, *34*, 27171–27183.
31. Li, Z.; Müller, T.; Evans, A.; Taylor, R.H.; Unberath, M.; Liu, M.Y.; Lin, C.H. Neuralangelo: High-fidelity neural surface reconstruction. In Proceedings of the IEEE/CVF Conference on Computer Vision and Pattern Recognition, Vancouver, BC, Canada, 17–24 June 2023; pp. 8456–8465.
32. Müller, T.; Evans, A.; Schied, C.; Keller, A. Instant neural graphics primitives with a multiresolution hash encoding. *ACM Trans. Graph. (TOG)* **2022**, *41*, 1–15. [CrossRef]
33. Sun, C.; Sun, M.; Chen, H.T. Direct voxel grid optimization: Super-fast convergence for radiance fields reconstruction. In Proceedings of the IEEE/CVF Conference on Computer Vision and Pattern Recognition, New Orleans, LA, USA, 18–24 June 2022; pp. 5459–5469.
34. Chen, A.; Xu, Z.; Geiger, A.; Yu, J.; Su, H. Tensorf: Tensorial radiance fields. In Proceedings of the 17th European Conference on Computer Vision, Tel Aviv, Israel, 23–27 October 2022; Springer: Cham, Switzerland, 2022; pp. 333–350.
35. Wu, T.; Wang, J.; Pan, X.; Xu, X.; Theobalt, C.; Liu, Z.; Lin, D. Voxurf: Voxel-based Efficient and Accurate Neural Surface Reconstruction. In Proceedings of the International Conference on Learning Representations (ICLR), Kigali, Rwanda, 1–5 May 2023.
36. Shen, J.; Ruiz, A.; Agudo, A.; Moreno-Noguer, F. Stochastic neural radiance fields: Quantifying uncertainty in implicit 3D representations. In Proceedings of the 2021 International Conference on 3D Vision (3DV), Virtual, 1–3 December 2021; pp. 972–981.
37. Shen, J.; Agudo, A.; Moreno-Noguer, F.; Ruiz, A. Conditional-flow nerf: Accurate 3d modelling with reliable uncertainty quantification. In Proceedings of the 17th European Conference on Computer Vision, Tel Aviv, Israel, 23–27 October 2022; Springer: Cham, Switzerland, 2022; pp. 540–557.
38. Shen, J.; Ren, R.; Ruiz, A.; Moreno-Noguer, F. Estimating 3D uncertainty field: Quantifying uncertainty for neural radiance fields. In Proceedings of the 2024 IEEE International Conference on Robotics and Automation (ICRA), Yokohama, Japan, 13–17 May 2024; pp. 2375–2381.
39. Blei, D.M.; Kucukelbir, A.; McAuliffe, J.D. Variational inference: A review for statisticians. *J. Am. Stat. Assoc.* **2017**, *112*, 859–877. [CrossRef]
40. Pan, X.; Lai, Z.; Song, S.; Huang, G. Activenerf: Learning where to see with uncertainty estimation. In Proceedings of the 17th European Conference on Computer Vision, Tel Aviv, Israel, 23–27 October 2022; Springer: Cham, Switzerland, 2022; pp. 230–246.
41. Martin-Brualla, R.; Radwan, N.; Sajjadi, M.S.; Barron, J.T.; Dosovitskiy, A.; Duckworth, D. Nerf in the wild: Neural radiance fields for unconstrained photo collections. In Proceedings of the IEEE/CVF Conference on Computer Vision and Pattern Recognition, Nashville, TN, USA, 20–25 June 2021; pp. 7210–7219.
42. Sünderhauf, N.; Abou-Chakra, J.; Miller, D. Density-aware nerf ensembles: Quantifying predictive uncertainty in neural radiance fields. In Proceedings of the 2023 IEEE International Conference on Robotics and Automation (ICRA), London, UK, 29 May–2 June 2023; pp. 9370–9376.
43. Lakshminarayanan, B.; Pritzel, A.; Blundell, C. Simple and scalable predictive uncertainty estimation using deep ensembles. *Adv. Neural Inf. Process. Syst.* **2017**, *30*, 6402–6413.
44. Zhan, H.; Zheng, J.; Xu, Y.; Reid, I.; Rezatofighi, H. Activermap: Radiance field for active mapping and planning. *arXiv* **2022**, arXiv:2211.12656.
45. Kajiya, J.T.; Von Herzen, B.P. Ray tracing volume densities. *ACM SIGGRAPH Comput. Graph.* **1984**, *18*, 165–174. [CrossRef]
46. Max, N. Optical models for direct volume rendering. *IEEE Trans. Vis. Comput. Graph.* **1995**, *1*, 99–108. [CrossRef]
47. Kingma, D.; Ba, J. Adam: A Method for Stochastic Optimization. In Proceedings of the International Conference on Learning Representations (ICLR), San Diego, CA, USA, 7–9 May 2015.
48. Sundermeyer, M.; Mousavian, A.; Triebel, R.; Fox, D. Contact-graspnet: Efficient 6-dof grasp generation in cluttered scenes. In Proceedings of the 2021 IEEE International Conference on Robotics and Automation (ICRA), Xi'an, China, 30 May–5 June 2021; pp. 13438–13444.
49. Ke, L.; Ye, M.; Danelljan, M.; Liu, Y.; Tai, Y.W.; Tang, C.K.; Yu, F. Segment anything in high quality. In Proceedings of the 37th International Conference on Neural Information Processing Systems, New Orleans, LA, USA, 10–16 December 2023; pp. 29914–29934.
50. Paszke, A.; Gross, S.; Massa, F.; Lerer, A.; Bradbury, J.; Chanan, G.; Killeen, T.; Lin, Z.; Gimelshein, N.; Antiga, L.; et al. PyTorch: An imperative style, high-performance deep learning library. In Proceedings of the 33rd International Conference on Neural Information Processing Systems, Vancouver, BC, Canada, 8–14 December 2019; pp. 8026–8037.
51. Hess, R. *Blender Foundations: The Essential Guide to Learning Blender 2.5*; Routledge: London, UK, 2013.

52. Dosovitskiy, A.; Beyer, L.; Kolesnikov, A.; Weissenborn, D.; Zhai, X.; Unterthiner, T.; Dehghani, M.; Minderer, M.; Heigold, G.; Gelly, S.; et al. An Image is Worth 16 × 16 Words: Transformers for Image Recognition at Scale. In Proceedings of the 9th International Conference on Learning Representations (ICLR), Virtual, 3–7 May 2021.
53. Kuffner, J.J.; LaValle, S.M. RRT-connect: An efficient approach to single-query path planning. In Proceedings of the 2000 ICRA. Millennium Conference. IEEE International Conference on Robotics and Automation, San Francisco, CA, USA, 24–28 April 2000; Volume 2, pp. 995–1001.
54. Olson, E. AprilTag: A robust and flexible visual fiducial system. In Proceedings of the 2011 IEEE International Conference on Robotics and Automation, Shanghai, China, 9–13 May 2011; pp. 3400–3407.
55. Wen, B.; Yang, W.; Kautz, J.; Birchfield, S. Foundationpose: Unified 6d pose estimation and tracking of novel objects. In Proceedings of the IEEE/CVF Conference on Computer Vision and Pattern Recognition, Seattle, WA, USA, 17–21 June 2024; pp. 17868–17879.
56. You, Y.; Xiong, K.; Yang, Z.; Huang, Z.; Zhou, J.; Shi, R.; Fang, Z.; Harley, A.W.; Guibas, L.; Lu, C. PACE: A Large-Scale Dataset with Pose Annotations in Cluttered Environments. In Proceedings of the 18th European Conference on Computer Vision, Milan, Italy, 29 September–4 October 2024; Springer: Cham, Switzerland, 2025; pp. 473–489.

applied sciences

MDPI

Article

Cost-Effective Localization of Mobile Robots Using Ultrasound Beacons and Differential Time-of-Flight Measurement

Basil Mohammed Al-Hadithi [1,2,*] and Carlos Pastor [1]

1 Intelligent Control Group, Centre for Automation and Robotics UPM–CSIC, Universidad Politécnica de Madrid, C/J. Gutiérrez Abascal, 2, 28006 Madrid, Spain; carlos.pastor.ramirez@alumnos.upm.es
2 Department of Electrical, Electronics, Control Engineering and Applied Physics, School of Industrial Design and Engineering, Universidad Politécnica de Madrid, C/Ronda de Valencia, 3, 28012 Madrid, Spain
* Correspondence: basil.alhadithi@upm.es

Abstract: This paper presents an innovative and cost-effective solution for the absolute localization of mobile robots using ultrasound beacons. The proposed system addresses the challenge of precise positioning within a controlled environment by employing Differential Time-of-Flight (ToF) measurements to determine the relative distances between the robot and optimally placed beacons. Unlike other ToF methods that require synchronization pulses, the proposed approach eliminates this requirement, significantly simplifying the setup and reducing system complexity. Furthermore, the system achieves a higher sampling rate than conventional synchronization-based systems, enhancing real-time performance. Detailed analysis and simulation demonstrate the system's ability to provide accurate and reliable localization. The results highlight the potential for broad application in various robotic environments, offering a robust solution for absolute positioning without complex synchronization strategies. This work underscores the advantages of using ToF measurements with ultrasound beacons and contributes to the ongoing development of efficient and cost-effective robotic localization systems.

Keywords: mobile robot localization; ultrasound beacons; time-of-flight (ToF) measurement; cost-effective robotics; real-time localization; position estimation; absolute positioning

1. Introduction

Accurately determining the position of a mobile robot is crucial to its effective operation. As such, robot localization has been the subject of extensive research since the inception of robotics. Active beacon navigation is a reliable positioning solution that is often implemented in navigation systems for ships, airplanes, and commercial mobile robots. Active beacons with known fixed positions serve as absolute reference points, ensuring highly accurate positioning information is obtained. As a result, this approach offers high positioning reliability.

Active beacon technologies are generally based on Time of Flight (ToF) and Angle of Arrival (AoA). ToF consists of precisely measuring the time that a signal travels between an emitter and a receiver. Using the speed of the signal through the medium, the total distance travelled by the signal can be calculated. This distance, combined with the known position of the emitter, can be used to determine the position of the receiver using trilateration or multilateration techniques. This method is used in [1,2], employing regular ToF and round-trip ToF to calculate the position of the client, obtaining an accuracy of several centimeters. In [3], a ToA method is proposed that combines two different types of beacons that achieve a reduction in the overall frequency that the client needs to sense, allowing the use of much simpler sensors. AoA, on the other hand, uses the angle of the incoming signal relative to the receiver's orientation to estimate the direction of the emitter. By measuring the angle of arrival from multiple emitters with known positions, the receiver's position can be estimated using triangulation techniques.This method is used in [4] to obtain the position of each of the individuals of a swarm of unmanned aerial vehicles. In [5], an alternative method is proposed that utilizes the Angle of Arrival (AoA), which involves an initial estimation of the client's position to adjust the beam forming process, thereby achieving more accurate results.

Additionally, other solutions propose merging previous techniques to improve localization precision. In [6], a hybrid method is presented that combines the ToF and AoA, designed to be used in environments with a dense amount of obstacles that hinder the performance of simpler methods. In [7,8], a method is suggested that combines the ToF and AoA with an angle of departure (AoD)-based method.

Although geometry methods, like trilateration or multilateration techniques, are still more widely employed [9], deep learning algorithms are gaining increasing popularity for this purpose. In [10], a method is proposed that employs ToF with a fully connected deep neural network that estimates the position. On the other hand, an AoA-based technique using a shallow neural network is presented in [11].

Active beacons offer the advantage that the data required to estimate the robot's position do not need as much processing compared to other techniques, so low-power solutions are easier to implement in beacon-based systems. In addition, the beacons are fixed and the positions of the beacons relative to the installation are known. This enables every calculated position to be directly related to the installation itself, providing absolute positioning without the necessity for inference.

As the prevalence of the Internet of Things (IoT) and robotics continues to rise, the demand for accurate and reliable positioning information at a low cost becomes more important across various applications, including smart cities, industrial settings, public spaces, healthcare, and domestic applications. This has sparked an increasing interest within the research community in the development of lightweight, low-power positioning solutions. In [12,13], different approaches to the localization of nodes in wireless sensor networks are presented using triangulation to estimate the position of a node. Another current development in the field of indoor positioning is the development of Ultra-Wide-Band Technology (UWB). In [14], an approach to localization using trilateration is proposed using a UWB radar with an antenna array that supports trilateration for indoor short-range positioning. Other systems integrate other technologies, such as Bluetooth [15], Wi-Fi [10,16], or radio frequency identification (RFID) [17], although these may require specialized sensors or a thorough mapping of the environment to understand the spatial distribution of signal characteristics [3].

In this work, a solution that uses ToF measurements is presented to determine the relative distances between a robot and optimally placed ultrasound beacons, facilitating accurate absolute positioning within a predefined map. By eliminating the need for synchronization pulses, the system setup is significantly simplified, which reduces the overall complexity. Furthermore, the proposed approach improves real-time performance by achieving a higher sampling rate compared to conventional synchronization-based systems.

The remainder of this paper is structured as follows. In Section 2, a localization solution is presented, detailing the principles and implementation of the proposed ultrasound-based system. In Section 3, the implementation details are provided, including the hardware setup and the measurement results. Section 4 presents the comprehensive simulation results and analysis that demonstrate the effectiveness of the proposed method. In Section 5, a discussion of the results obtained and their implications in the field of mobile robot localization is provided. Finally, Section 6 concludes the paper, summarizing the key findings.

2. Localization Solution

This section introduces a novel and cost-effective localization solution that uses ultrasound technology to achieve precise positioning without complex synchronization mechanisms. The proposed system utilizes ultrasound beacons and receivers, coupled with an innovative time difference measurement approach, to determine a robot's position within a predefined area. The following subsections will detail the background of ultrasound technology in localization, the core principles of the proposed system, its base operation, the analysis of the localization strategy, and the mathematical principles underlying the approach.

Ultrasound is a commonly used technology for indoor positioning and beacon systems, which is notable due to its favourable cost/performance ratio, wide availability, and robustness in challenging environments. Various approaches to ultrasonic localization exist, demonstrating the technology's versatility and reliability in numerous industrial and commercial applications. For example, in [18], a relative positioning system is demonstrated using three ultrasonic sensors to obtain Time-of-Flight (ToF) data and using the time differences between the signals to calculate the robot coordinates. Another application in which ultrasound is widely used is obstacle detection and avoidance. In [19], a small autonomous mobile robot is presented equipped with a microcontroller and four ultrasonic sensors to avoid real-time obstacles, using a linear recursive Kalman filter to improve the response of the system and ensure smoother movement. These examples illustrate the range of applications for ultrasound technology, from presence detection to parking systems.

2.1. System Design, Core Innovations, and Benefits

Unlike traditional localization methods that rely on trilateration and require precise synchronization signals (such as infrared light), the system presented in this work operates on a fundamentally different principle. By eliminating the need for absolute ToF measurements and synchronization mechanisms, the system's design is significantly simplified while simultaneously enhancing accuracy and reliability. This approach differs from conventional beacon systems demonstrated in [20,21], which typically use ultrasound for ToF measurements in combination with synchronization signals.

The core innovation of this approach lies in its ability to determine position using only the time differences between received ultrasound signals. This method eliminates the need for synchronization between beacons and the robot. As a result, the solution offers substantial advantages in terms of reduced system complexity, lower implementation costs, and improved robustness against timing errors.

The proposed system is specifically tailored for real-time applications in dynamic indoor environments, such as hospitals, museums, and other public spaces with foot traffic. Thanks to ultrasound technology, the system maintains robustness in changing environments and supports multiple robots simultaneously, ensuring consistent performance where it is essential. Moreover, the installation cost is economical, as multiple robots can share the same ultrasound pulses.

A standout feature of our system is its relatively low computational footprint, especially compared to SLAM or vision-based localization methods. The simplicity of our approach could allow for real-time processing on a high-performance microcontroller, avoiding the need for more powerful and energy-intensive processors running a full OS. This not only reduces overall system costs but also significantly lowers power consumption.

The system comprises n multiple beacons and n receivers, optimally placed to cover the operational area. The network of ultrasound beacons emit signals at precise intervals and in a specific order, while receivers mounted on the mobile robot detect and process these signals for position calculation.

The system's ability to function without requiring precise timing synchronization allows for improved flexibility across diverse indoor settings. It can adjust more readily to different spatial configurations and is less susceptible to environmental factors that might disrupt precise timing signals. Furthermore, the simplification of the hardware requirements makes the system more cost-effective to implement and maintain, without sacrificing accuracy.

2.2. System Overview

Figure 1 illustrates the overall setup of our ultrasound-based localization system. Ultrasound beacons are mounted on the ceiling to provide consistent coverage and minimize obstructions. The mobile robot, equipped with ultrasound receivers, moves within the area covered by the beacons. This configuration allows for accurate positioning of robots with varying heights, as the system can account for the vertical offset between the beacons and the robot-mounted receivers.

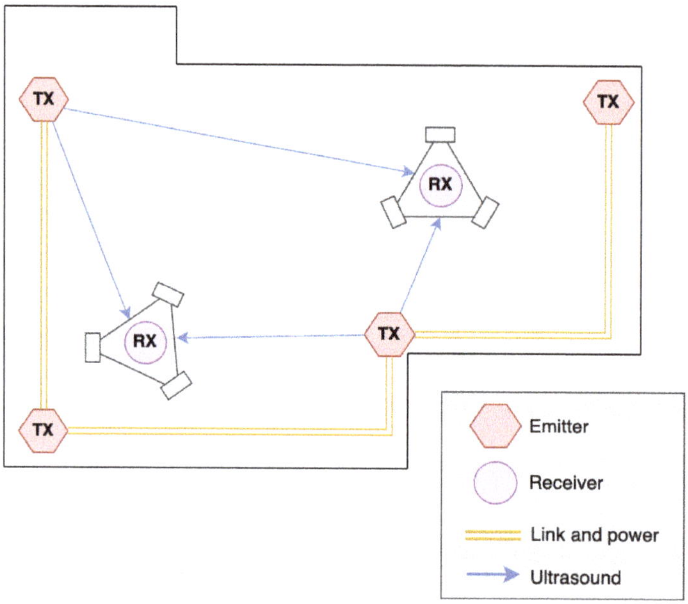

Figure 1. Overview of the ultrasound-based localization system, showing the relationship between ceiling-mounted beacons and the mobile-robot-mounted receivers.

2.3. Base Operation

The localization system employs a sequential ultrasound emission strategy in combination with a time difference measurement approach. This method circumvents the need for beacon–receiver synchronization typical in ToF systems. The operational sequence and signal processing technique are as follows:

Signal Emission: The system operates by firing the beacons in pairs according to a pre-determined sequence. Each pair of beacons emits an ultrasound signal simultaneously. This paired emission strategy allows for relative time measurements, reducing the complexity and potential errors associated with absolute time synchronization. By using pairs of beacons, the system can create distinct time differences that are crucial for the localization process.

Time Difference Measurement: As the signals travel from the beacons to the receivers, the system measures the time differences (Δt) between the arrivals of these signals at each receiver. These time differences are then used to calculate the relative distances between the beacons and the robot, based on the known speed of sound in the medium. By using time differences rather than absolute times, the need for precise clock synchronization between beacons and receivers is eliminated.

Position Calculation: By analyzing the time differences from multiple pairs of beacons, the system can triangulate the precise position of the robot. The calculation utilizes hyperbolic multilateration techniques, where each time difference measurement defines a hyperbolic curve on which the robot must lie. The intersection of these curves, derived from multiple beacon pairs, pinpoints the robot's position with high accuracy. This approach, based on the known positions of the beacons and the measured time differences, allows for high-accuracy localization without the need for complex synchronization mechanisms.

2.4. Working Principle

The core principle of this localization system revolves around measuring the time differences (Δt) between the arrivals of signals from pairs of beacons at the robot's receivers. These time differences can be interpreted as hyperbolic loci in space, known as isochrones (from Greek "iso-" meaning "equal" and "chronos" meaning "time"), which represent all points where the time difference between signals from two beacons remains constant.

All possible solutions for a given time difference lie on a hyperbolic function in space. The isochrones are a visualization of this hyperbolic space, represented as color gradients in the figures. By analyzing the intersections of these isochrones from multiple pairs of beacons, the system can determine the robot's precise location with high accuracy.

To better illustrate this concept, Figures 2 and 3 provide visual representations of the hyperbolic function through color gradients. Each color in the gradient corresponds to a specific time difference value, effectively representing the function across the operational area. These visualizations offer valuable insight into the geometric relationships that underpin the system's functionality and demonstrate how time differences translate into spatial locations.

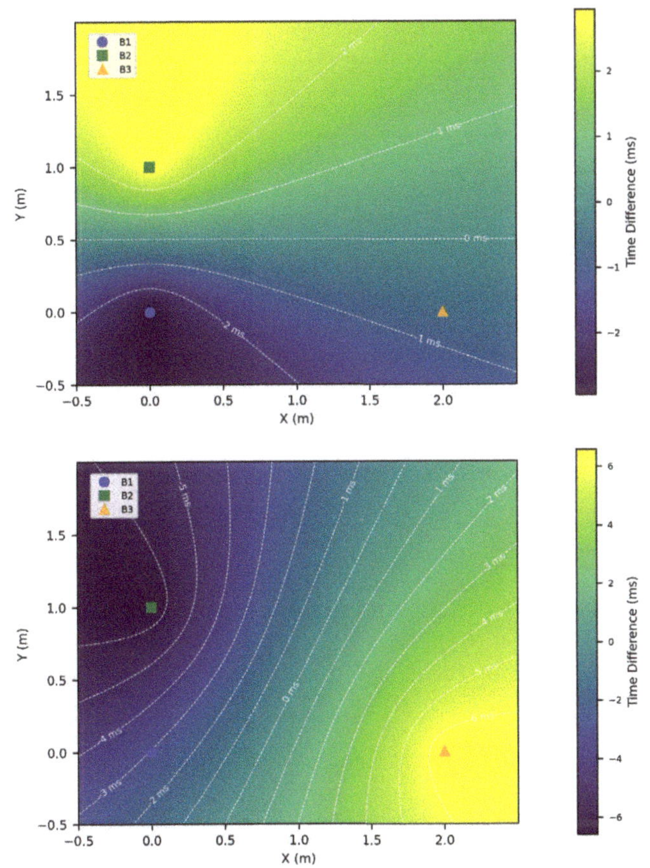

Figure 2. *Cont.*

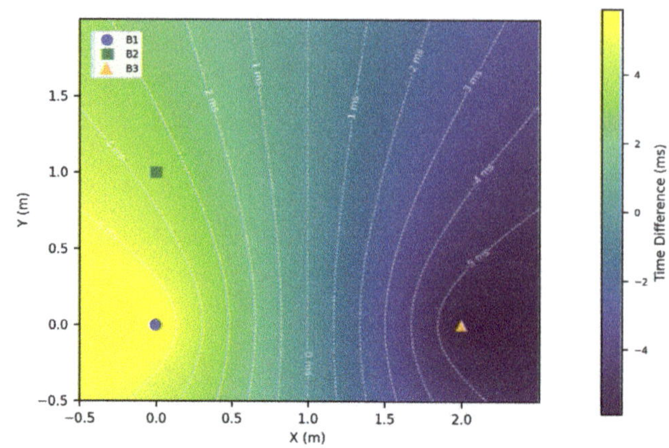

Figure 2. Time difference plots for pairs of beacons. The color gradients represent the hyperbolic function, with isochrones in (Δt) 1 millisecond increments. The x- and y-axes represent spatial coordinates within the operational area.

When a pair of beacons emit ultrasound signals simultaneously, the robot's receivers detect the arrival times of these signals. The system computes the time difference (Δt) between the signal arrivals, placing the robot on an isochrone that represents all possible locations where this time difference could be observed. The intersection of isochrones from multiple beacon pairs provides the robot's precise location.

Figure 3 shows the complete isochrone plot for all beacon pairs. Each color represents isochrones for a different beacon pair, illustrating how time differences vary across the operational area. The dense network of isochrones demonstrates the rich information available from just three beacons, enabling precise localization.

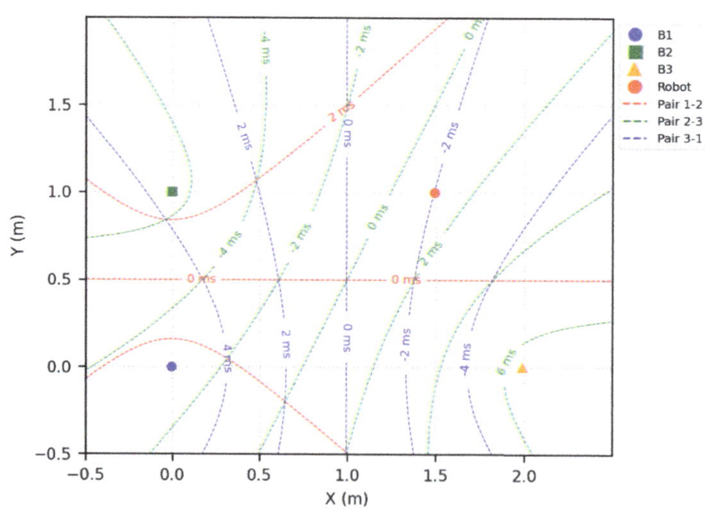

Figure 3. All isochrones for beacon pairs. Each color represents a different beacon pair, showing how time differences vary across the area.

Figure 4 focuses on the specific isochrones that pass through the robot's position. This visualization clearly shows how the intersection of these isochrones from different beacon pairs determines the robot's exact location. Each colored line represents the isochrone from a beacon pair that corresponds to the time difference measured by the robot.

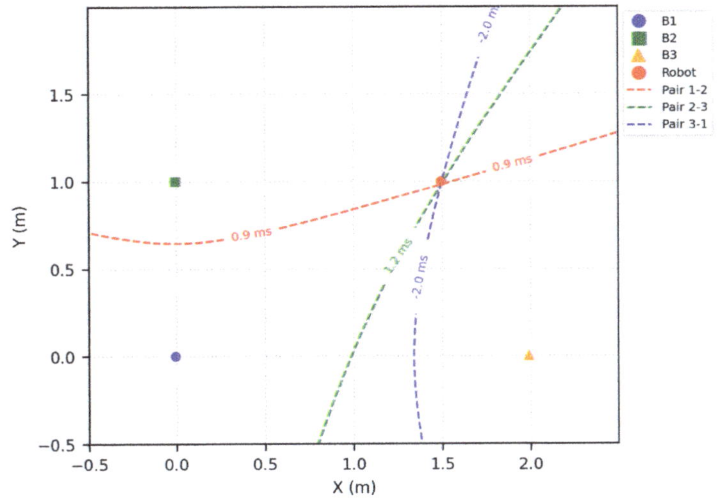

Figure 4. Isochrones passing through the robot's position. The intersection of these isochrones from different beacon pairs indicates the robot's precise location.

These visualizations highlight the effectiveness of the time difference approach. They demonstrate how this method achieves accurate localization by leveraging the geometric relationships between the beacons and the robot.

2.5. Mathematical Principles of Hyperbolic Multilateration

The core of the localization method developed in this work is based on the analysis of isochrone curves generated by time differences (Δt) between signals from pairs of beacons. These isochrones are hyperbolic curves that represent all points in space where the difference in the distances to two beacons is constant.

While this mathematical model forms the foundation of our localization system, the practical implementation involves considerations beyond the scope of this analysis, particularly regarding optimal beacon placement.

For a pair of beacons located at (x_1, y_1) and (x_2, y_2), the equation for the hyperbolic isochrone curve can be expressed as follows:

$$\sqrt{(x-x_1)^2+(y-y_1)^2}-\sqrt{(x-x_2)^2+(y-y_2)^2}=\Delta t\cdot v \tag{1}$$

where

- (x, y) represents any point on the hyperbola;
- Δt is the measured time difference between signal arrivals from the two beacons;
- v is the speed of sound in the medium (approximately 343 m/s in air at room temperature).

To determine the robot's position, the intersection of isochrone curves from different pairs of beacons should be found. Consider a system with three beacons located at (x_1, y_1), (x_2, y_2), and (x_3, y_3). Two hyperbolic equations can be formed:

$$\begin{cases}\sqrt{(x-x_1)^2+(y-y_1)^2}-\sqrt{(x-x_2)^2+(y-y_2)^2}=\Delta t_{12}\cdot v\\ \sqrt{(x-x_2)^2+(y-y_2)^2}-\sqrt{(x-x_3)^2+(y-y_3)^2}=\Delta t_{23}\cdot v\end{cases} \tag{2}$$

where Δt_{12} and Δt_{23} are the measured time differences between signals from beacons 1 and 2, and beacons 2 and 3, respectively.

Solving these nonlinear equations simultaneously provides the coordinates (x, y) of the robot. This solution represents the unique point where all the measured time differences are satisfied, thus pinpointing the robot's position.

2.6. Adapting the Model for Ceiling-Mounted Beacons

To ensure the localization system's practicality and wide applicability, ceiling-mounted ultrasound beacons were chosen to support ground-based robots of varying heights. This vertical separation between beacons and receivers introduces a modification to the modeling that requires the addition of a plane offset to the mathematical model. This adaptation is necessary for the system's real-world functionality, enabling accurate localization in three-dimensional space while maintaining the system's core principles.

Mathematical Adaptation

To incorporate this ceiling-to-ground configuration and variable robot heights, the equation must be modified as follows:

$$\sqrt{(x-x_i)^2+(y-y_i)^2+(H-h_r-z_i)^2}-\sqrt{(x-x_j)^2+(y-y_j)^2+(H-h_r-z_j)^2}=\Delta t_{ij}\cdot v \quad (3)$$

where

- $(x, y, 0)$ represents the robot's position on the ground plane;
- (x_i, y_i, z_i) and (x_j, y_j, z_j) are the coordinates of two ceiling-mounted beacons;
- H is the ceiling height;
- h_r is the height of the specific robot;
- Δt_{ij} is the measured time difference between signal arrivals from beacons i and j;
- v is the speed of sound in the medium.

This equation ensures that the calculated position accurately reflects the robot's true location on the ground, accounting for its specific height and the ceiling-mounted beacon configuration.

2.7. Considerations on Beacon Placement

The optimal placement of beacons in a given environment is a complex problem that warrants its own dedicated study. It depends on various factors specific to each implementation scenario, including room geometry, potential obstructions, expected robot paths, and desired coverage area. While a comprehensive analysis of optimal beacon placement is beyond the scope of this paper, it is important to note that our system's flexibility allows for beacon positions to be optimized during the installation phase. This adaptability enables the system to be fine-tuned to best suit the particular requirements and constraints of each environment. Factors that might influence beacon placement include the following:

- Maximizing coverage area;
- Minimizing geometric dilution of precision;
- Avoiding areas of potential signal interference;
- Accommodating architectural features or obstacles.

The process of determining optimal beacon placement could involve techniques such as computer simulations, on-site testing, or even machine learning algorithms. However, such optimization methods represent a separate field of study and are not explored in detail in this work. Our focus remains on the fundamental principles and advantages of the proposed localization system, with the understanding that practical implementations would involve additional considerations for beacon placement.

3. Implementation Details

The performance and efficacy of the proposed ultrasound-based localization system were evaluated through a combination of hardware implementation, practical measurements, and simulation. The results demonstrate the system's capability to accurately measure time differences between ultrasound signals, which is crucial for precise localization.

3.1. Hardware

A test system is developed to validate the idea principle and take measurements to evaluate possible performance and limitations.

The amplifier board, shown in Figure 5, is designed with a high-impedance operational amplifier set to offset the signal and provide gain, allowing both sides of the oscillation to be viewed within the 0 V to 3.3 V range, making it compatible with a microcontroller ADC. The gain can be fine-tuned using an on-board electronic potentiometer, allowing for sensitivity control during operation. This sensor board measures 3 cm by 3 cm.

Figure 5. Amplifier board from the top and bottom sides, respectively.

The breakout board, shown in Figure 6, is used solely for testing purposes. It facilitates the connections necessary to sample the data during the experiments. This board measures 5 cm by 4 cm.

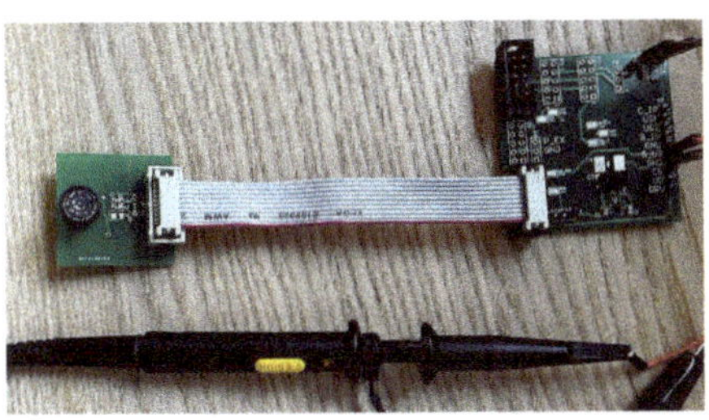

Figure 6. Breakout board used for sampling data during testing.

The emitters, depicted in Figure 7, are based on an STM Nucleo-32 board and generate pulses to drive the ultrasound emitter. An ultrasound driver is implemented using an H-bridge configuration, which allows for efficient switching of the voltage polarity to the ultrasound transducer, thereby maximizing the output power. The width and number of pulses can be adjusted via software. The boards are connected using flat cables to form the emitter network. The emitter device measures 10 cm long by 7 cm tall by 6 cm wide.

Figure 7. Emitter device used for testing, based on an STM Nucleo-32 board.

The hardware was developed with the purpose of demonstrating the viability and capability of capturing precise time differences in ultrasound signal arrivals with a very low-cost approach. It is important to note that these dimensions represent the current development and testing hardware. In a final implementation, the hardware would be significantly miniaturized. The current larger sizes are due to the use of development boards, connectors, and other components that facilitate testing and modification. A production version would be optimized for size and efficiency, resulting in much more compact hardware suitable for integration into various robotic platforms.

3.2. Test Design

The experimental design for this study was chosen to demonstrate the fundamental viability of our proposed ultrasound-based localization system. Our experiments focused on three key aspects:

1. Static accuracy: Single-point and two-point measurements were conducted to establish the system's baseline accuracy and its ability to distinguish between different positions.
2. Dynamic tracking: A circular motion experiment was performed to evaluate the system's performance in tracking a moving target.
3. Time difference precision: Measurements with varying distances between emitters were taken to evaluate the system's general precision and sensitivity.

While these tests are limited in scope, they serve as a proof of concept, providing concrete evidence of the system's functionality and potential.

3.3. Measurement

To test the capture system, two sets of measurements were taken with the emitters positioned 0.5 m and 1 m apart from each other, with the closest sensor being 1 m from the source using the developed sensor. These captures demonstrate that the system is capable of precisely detecting small differences in time. However, it should be noted that when the time differences are very small, the overlapping of waves can occur, making it challenging to distinguish between them.

Figure 8 presents the oscilloscope capture with emitters positioned 1000 mm and 1500 mm from the receiver. The system's precision is demonstrated by the clearly distinct arrival times of the signals and their subsequent attenuation.

To further evaluate the system's repeatability, Figure 9 depicts the arrival and attenuation of the signals with one emitter 1000 mm and the other 2000 mm from the receiver, averaged over 100 samples. This averaging demonstrates minimal jitter, underscoring the high repeatability of the system.

The gain of the amplifier was set at a level adequate for direct reception, which minimizes the impact of echoes. This configuration ensures that echoes are very small compared to the direct signal, enhancing the system's accuracy in detecting the primary signals.

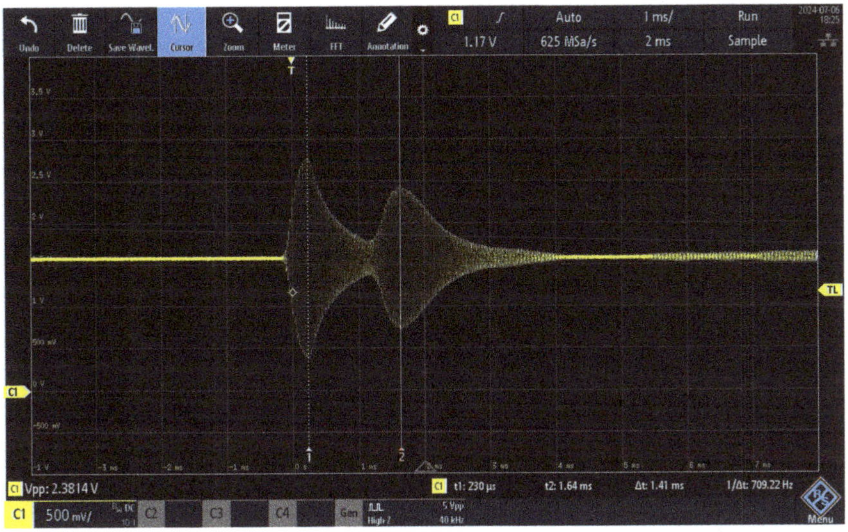

Figure 8. Oscilloscope capture showing the arrival of two ultrasound signals with one emitter placed 1000 mm and the other one 1500 mm from the receiver, highlighting the signal's initial peak and attenuation over time.

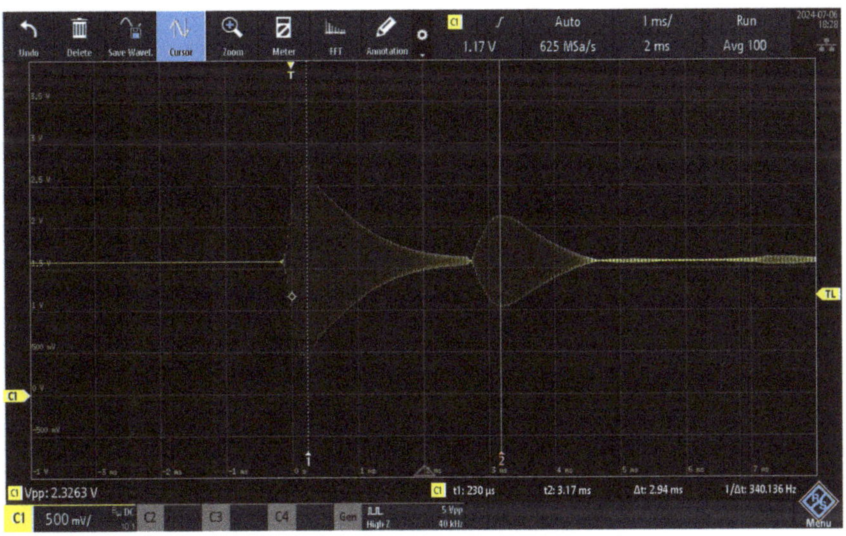

Figure 9. Oscilloscope capture depicting the arrival and attenuation of two ultrasound signals with one emitter 1000 mm and the other one 2000 mm from the receiver, averaged over 100 samples to demonstrate minimal jitter.

These measurements demonstrate the effectiveness of the developed sensor in precisely capturing small time differences between ultrasound signals, despite the challenge of wave overlap at very short time intervals. The high directivity of the sensor effectively mitigates the problem of small intervals. Placing an array of three or four sensors per robot resolves this issue and increases accuracy. The high precision of the amplifier and the system's repeatability are evident, confirming its capability for accurate and reliable time difference measurements.

4. Simulation Results and Analysis

In this section, we present the results of our simulation, focusing on the relative error and identification of critical areas in a sample system with real sampling times. Our initial testing revealed that the timing error was extremely low, so this error source was discarded from the analysis. To validate and evaluate the performance of the system, we performed a simulation comparing the localization accuracy across various scenarios. We introduced normalized errors in the sample times, speed of sound, and measurement accuracy based on data collected during our testing. This approach allowed us to more accurately model real-world conditions and assess the system's robustness in the face of typical variations and uncertainties.

4.1. Static Localization Performance

Figure 10 provides a comprehensive overview of our localization system's performance across the entire test area. Key observations are as follows.

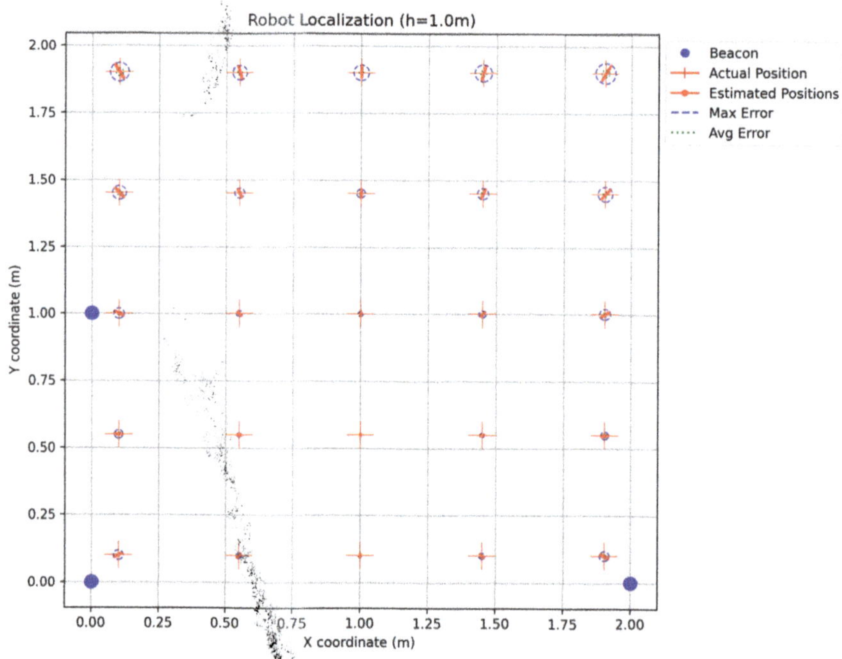

Figure 10. Localization results for robot height 1.0 m. The red crosses represent actual positions; red dots are estimated positions. Blue dashed circles show maximum error and green dotted circles show average error. Blue circles indicate beacon positions at (0, 0), (2, 0), and (1, 0).

The linear arrangement of beacons along the x-axis significantly influences the system's performance, creating a unique error pattern. Estimated positions (red dots) cluster around actual positions (thin red crosses), visually indicating the system's precision. Error circles (blue dashed for maximum, green dotted for average) vary in size across the test area, suggesting non-uniform accuracy. Accuracy appears to be higher near the center of the beacon arrangement and decreases towards the edges of the covered area, particularly in the y-direction. There is a slight elongation of error circles in the y-direction, particularly for points further from the x-axis, likely due to the linear beacon arrangement.

4.2. Analysis of a Central Point

Figure 11 offers a zoomed-in view of a single test point at the center of our grid (1.0 m, 1.0 m), allowing for a detailed examination. The distribution of estimates (red dots) forms a line around the actual position, indicating balanced errors in the x- and y-directions at this central location. The tight clustering of estimates suggests high precision, with most estimates falling within a small area. The proximity of the estimated cluster to the actual position (red cross) demonstrates high accuracy with no significant systematic bias. The relatively small size of the error circles, particularly the average error circle (green dotted), indicates good overall accuracy at this central point. This central point's equidistant position from all beacons likely contributes to its high accuracy, showcasing the optimal performance of the system under ideal conditions.

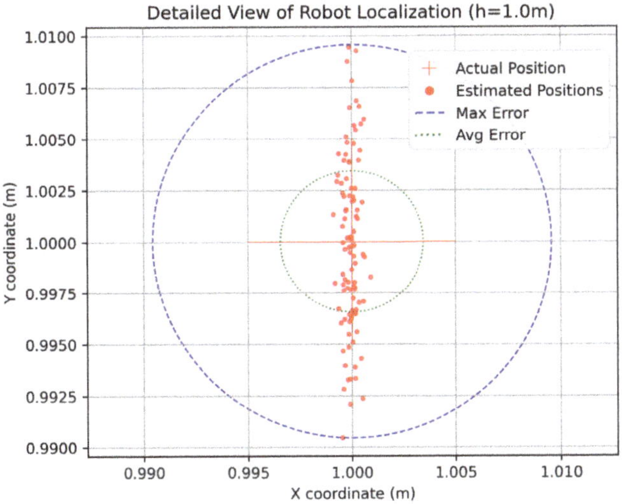

Figure 11. Detailed view of localization results for a single position at (1.0 m, 1.0 m). The thin red cross represents the actual position. Red dots are estimated positions. The blue dashed circle shows the maximum error, and the green dotted circle shows the average error.

4.3. Error Distribution Analysis

Figure 12 presents the comprehensive error distribution across all test points and estimates. The histogram reveals a right-skewed distribution, typical for error measurements in localization systems. The mean error (red dashed line) is slightly larger than the median error (green dotted line), consistent with the right-skewed nature of the distribution. This suggests that while most errors are small, occasional larger errors pull the mean higher. Most errors are concentrated in the left part of the distribution, indicating that the majority of estimates have relatively small errors, which is a positive indicator of the system's overall accuracy. The long tail to the right shows that larger errors do occur, albeit less frequently. These could represent estimates at the edges of our test area, where the system's performance is expected to degrade. The peak of the distribution occurs at a low error value, further confirming the system's overall accuracy for most measurements.

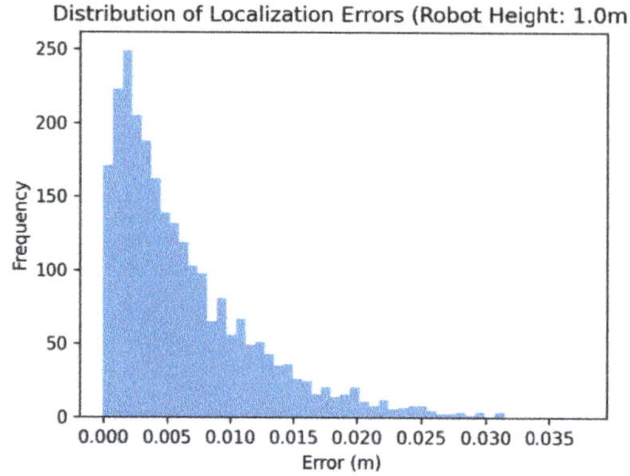

Figure 12. Distribution of localization errors. Blue histogram bars show error frequency.

4.4. Dynamic Localization Performance

Figure 13 illustrates the system's performance in tracking a robot moving in a circular path. The actual path (blue line) forms a perfect circle, centered at (1, 1) with a radius of 0.5 m. Estimated positions (colored dots) generally follow the circular path, demonstrating the system's ability to track dynamic movement. The color gradient of the estimated positions provides insight into the error distribution during motion. Lighter colors (lower errors) seem to dominate, indicating generally good tracking performance. Some areas of the circle show darker colored estimates, suggesting higher errors. These could be points where the robot's motion aligns unfavorably with the beacon arrangement. The beacons (blue dots at the corners) provide context for understanding how their positioning affects the tracking accuracy at different points along the path.

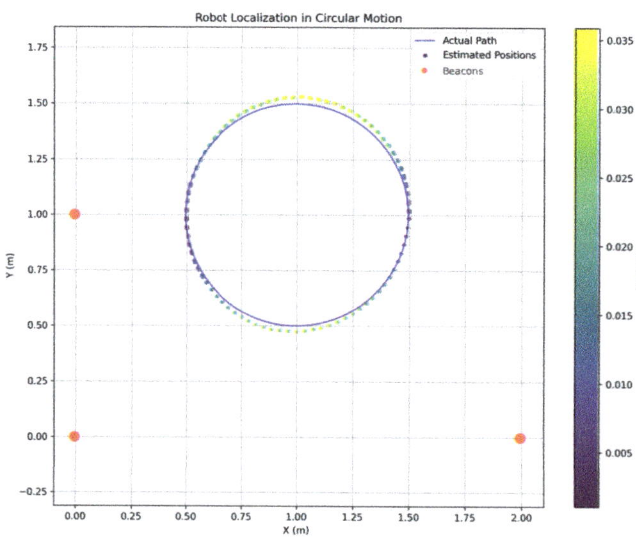

Figure 13. Robot localization during circular motion. The blue line represents the actual path, while colored dots show estimated positions. The color gradient indicates the magnitude of error, with the yellow color representing larger errors and the blue color smaller errors. Red circles indicate beacon positions at (0, 0), (2, 0), and (1, 0).

Figure 14 complements the circular motion analysis by showing how the localization error varies with the position of the robot. The error plot shows a periodic pattern, corresponding to the robot's revolutions around the circular path. The shape of the measured circle is due to the robot changing position slightly between measurements.

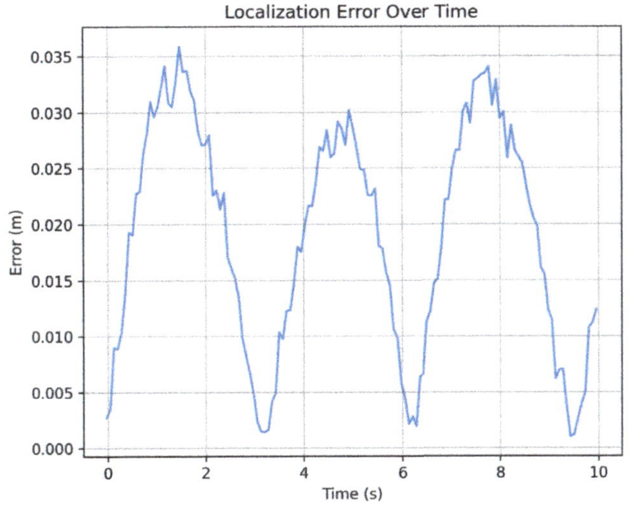

Figure 14. Localization error over time during circular motion. The graph shows how the error varies as the robot completes multiple revolutions.

5. Discussion

This section discusses the implementation, advantages, limitations, and implications of the proposed ultrasound-based localization system for mobile robots. It also compares this approach to existing technologies.

5.1. Implementation Overview

The localization system utilizes ceiling-mounted ultrasound beacons and accommodates robots of varying heights:

- Beacons are mounted on the ceiling to provide consistent coverage and minimize obstructions.
- The system is designed to work with robots of different heights, enhancing its versatility.
- The plane offset between beacons and robots varies based on robot and ceiling height.

5.2. Key Advantages

The proposed system offers several advantages:

- Cost-effectiveness: Ultrasound components are inexpensive compared to other localization technologies, making the system more affordable for large-scale deployments.
- Environmental robustness: The system performs well in various lighting conditions, unlike optical or laser-based methods.
- Differential measurement: By focusing on time differences rather than absolute times, this approach mitigates variations due to environmental factors.
- Scalability: The system can be easily expanded by adding more beacons, allowing for coverage of larger areas or improved accuracy.

5.3. Limitations and Challenges

While our system offers numerous benefits, it is important to acknowledge its limitations and potential challenges:

- Range limitations: Ultrasound signals attenuate relatively quickly in air, potentially requiring a higher density of beacons for larger areas. However, for the applications we envision—installations with foot traffic and dynamic environments—the current range of 6 to 10 m should cover all cases.
- Acoustic interference: Environments with significant ambient noise or other ultrasonic sources may experience interference. However, the ultrasound band is relatively quiet, and the narrow frequency of our emitted signal allows for easy filtering, reducing potential ambient interference.
- Update rate: The system's update rate may be lower compared to some other technologies. For the target application with robots that move at walking pace this is less of a concern.
- Obstacles: Solid obstacles between beacons and receivers could potentially block signals. However, ultrasound can pass around obstacles with some power loss, and our system has sufficient gain to detect very soft signals. Additionally, placing the receiver high on the robot will mitigate most potential signal blockages.

It is worth noting that these limitations are less significant in our target application of indoor environments with walking-speed robots. In such settings, the update rate is sufficient, range limitations are manageable, and acoustic interference can often be minimized through environmental design. The installation of a solution is also part of the system, and any coverage limitations should be addressed in this phase.

5.4. Comparison with Other Positioning Technologies

To better understand the strengths and limitations of the proposed ultrasound-based localization system, it is compared with other common positioning technologies: vision-based systems, LiDAR, SLAM, and traditional odometry. Table 1 provides a comprehensive comparison of these technologies across key performance metrics, including localization accuracy, scale, system cost, sensor cost, and adaptability.

Table 1. Comparison of localization technologies.

Technology	Accuracy	Scale	System Cost	Sensor Cost	Adaptability
US Beacon	Medium	Room-level	Low	Low	High
Vision SLAM	Medium	Room-level	Medium	Medium	Low–Medium
LiDAR SLAM	High	Medium-scale	High	High	Low–Medium
Odometry	Low	Medium-scale	Low	Low	High

Vision-based localization systems have gained popularity due to their ability to provide rich environmental data. However, these systems often struggle in low-light conditions or environments with limited visual features. In contrast, the ultrasound system operates independently of the lighting conditions, making it more versatile in various indoor settings. Furthermore, the ultrasound system generally requires less computational power compared to complex image processing algorithms, offering a more cost-effective solution. This lower processing requirement is a significant advantage for applications where power consumption is a concern.

LiDAR (Light Detection and Ranging) is known for its high accuracy, fast update rates, and excellent range and resolution, which makes it well suited for outdoor or large-scale indoor applications. Although LiDAR provides superior performance in these aspects, the ultrasound system is generally more cost-effective, particularly for smaller-scale applications, and offers absolute positioning in real time without the need for additional algorithms. The ultrasound system typically consumes less power than LiDAR. However, LiDAR's superior range and resolution make it more suitable for applications requiring detailed environmental mapping or long-range sensing. Simultaneous Localization and Mapping (SLAM) has become increasingly popular, especially with the availability of cost-effective sensors. While SLAM offers flexibility in unknown environments, it may be less reliable in dynamic settings compared to the proposed ultrasound system. The beacon-based system generally has lower computational needs compared to most SLAM implementations, allowing for simpler, more power-efficient hardware. Furthermore, the proposed system provides absolute positioning within the beacon network, while SLAM typically provides relative positioning that may drift over time without loop closure. The ultrasound-based approach can be more robust in dynamic environments where the physical space is changing, which can challenge SLAM algorithms.

Traditional odometry, often using wheel encoders or Inertial Measurement Units (IMUs), offers a cost-competitive alternative for robot localization. Like the ultrasound beacon-based system, odometry solutions can be implemented with relatively inexpensive hardware and have minimal computational requirements. However, odometry suffers from cumulative errors over time, leading to a drift in position estimates. This is particularly problematic for long-term operation or after sharp turns. In contrast, the ultrasound-based system provides absolute positioning that does not degrade with time. While odometry can offer very high update rates, making it suitable for real-time control of fast-moving robots, it often needs to be combined with other localization methods for reliable long-term operation.

Recent advances in SLAM and other advanced positioning strategies have made significant progress, particularly with the increasing availability of cost-effective LiDAR and camera solutions. However, these technologies still face challenges in dynamic environments and in optimizing cost and power consumption for certain applications. In contrast, the proposed ultrasound-based system offers a cost-effective alternative, as the main investment is made in the installation of the beacons, while the robots only require low-cost receivers. This is particularly advantageous for scenarios involving a fleet of robots, where equipping each robot with expensive sensors can be cost-prohibitive.

Despite its limitations in sampling rate and range compared to some other technologies, the ultrasound-based system provides a balanced solution that is especially suitable for controlled indoor environments. In these settings, factors such as cost-effectiveness, robustness, and simplicity often take precedence over high-speed tracking or large-scale mapping capabilities. The proposed system's ability to provide accurate, absolute positioning with minimal computational overhead and power consumption makes it an attractive choice for a wide range of indoor applications.

Ultimately, the choice between these technologies depends on the specific requirements of the application. Factors to consider include the type of environment (e.g., indoor, outdoor, and dynamic), accuracy requirements, update rate needs, available computational resources, power constraints, and budget. The ultrasound-based isochrone method presented in this paper offers a unique combination of benefits that make it particularly well suited for indoor applications demanding accurate and efficient localization.

By leveraging the advantages of ultrasound technology, such as its robustness to lighting conditions and low computational requirements, the proposed system fills a crucial gap in the spectrum of localization solutions. It provides a cost-effective and reliable alternative for applications where the operating environment is well defined, and the primary focus is on achieving precise positioning without the need for complex, resource-intensive algorithms or expensive sensor suites.

5.5. Real-World Applications

The ultrasound-based localization system offers significant advantages in public spaces with high pedestrian traffic. Unlike SLAM or LiDAR systems, it maintains accuracy in dynamic environments, relatively unaffected by moving individuals or varying crowd densities. This robustness, combined with its non-reliance on visual data, makes it effective in areas such as retail stores, hospitals, and museums, where consistent performance with people traffic is essential for robotic operation.

The system is especially well suited for medical supply robots in hospital corridors, positioning interactive guide robots in museums, and other public spaces that offer a controlled environment where an installation can be made. In traditional robotics applications such as warehouse and factory automation, it can serve as a complementary system or cost-effective solution for precise absolute positioning of inventory management robots or automated guided vehicles (AGVs). Its capacity to provide accurate localization in the presence of moving objects or personnel makes it valuable in environments where human workers and automated systems coexist, offering a reliable alternative or backup to existing positioning methods.

6. Conclusions

In conclusion, the ultrasound-based localization system offers a unique combination of cost-effectiveness, environmental robustness, and scalability, which makes it particularly well suited for indoor robotics applications. Although it has limitations in terms of range and update rate, these are often outweighed by its advantages in many practical scenarios. As robot deployment in indoor environments continues to grow, this approach provides a valuable addition to the toolkit of localization technologies, especially for applications that prioritize affordability and simplicity without compromising accuracy.

This work contributes to the field of robot localization by demonstrating the viability of a low-cost, computationally efficient approach to absolute positioning in indoor environments. This challenges the notion that high-precision indoor localization necessarily requires expensive sensors or complex algorithms. Furthermore, the use of differential measurements provides a novel approach to mitigating environmental variations in ultrasound-based systems.

Author Contributions: Conceptualization, B.M.A.-H. and C.P.; methodology, B.M.A.-H. and C.P.; software, C.P.; validation, B.M.A.-H. and C.P.; formal analysis, B.M.A.-H. and C.P.; investigation, B.M.A.-H. and C.P.; resources, B.M.A.-H. and C.P.; data curation, C.P.; original draft preparation, C.P.; review and editing, B.M.A.-H.; visualization, C.P.; supervision, B.M.A.-H. All authors have read and agreed to the published version of the manuscript.

Funding: This paper is part of the R&D project "Cognitive Personal Assistance for Social Environments (ACOGES)", reference PID2020-113096RB-I00, funded by MCIN/AEI/10.13039/501100011033.

Institutional Review Board Statement: Not applicable.

Informed Consent Statement: Not applicable.

Data Availability Statement: The original contributions presented in the study are included in the article, further inquiries can be directed to the corresponding author.

Conflicts of Interest: The authors declare no conflicts of interest.

References

1. Sukop, M.; Grytsiv, M.; Jánoš, R.; Semjon, J. Simple Ultrasonic-Based Localization System for Mobile Robots. *Appl. Sci.* **2024**, *14*, 3625. [CrossRef]
2. Maletic, N.; Sark, V.; Ehrig, M.; Gutiérrez, J.; Grass, E. Experimental evaluation of round-trip ToF-based localization in the 60 GHz band. In Proceedings of the 2019 International Conference on Indoor Positioning and Indoor Navigation (IPIN), Pisa, Italy, 30 September–3 October 2019; IEEE: Piscataway, NJ, USA, 2019; pp. 1–6.
3. Lin, Q.; An, Z.; Yang, L. Rebooting ultrasonic positioning systems for ultrasound-incapable smart devices. In Proceedings of the 25th Annual International Conference on Mobile Computing and Networking, Los Cabos, Mexico, 21–25 October 2019; pp. 1–16.
4. Hu, Y.; Jiang, Y.; Liu, Y.; He, X. Autonomous Control System with Passive Positioning for Unmanned-Aerial-Vehicle-Assisted Edge Communication in 6G. *Appl. Sci.* **2023**, *13*, 11014. [CrossRef]
5. Fascista, A.; Coluccia, A.; Wymeersch, H.; Seco-Granados, G. Low-complexity accurate mmwave positioning for single-antenna users based on angle-of-departure and adaptive beamforming. In Proceedings of the ICASSP 2020-2020 IEEE International Conference on Acoustics, Speech and Signal Processing (ICASSP), Barcelona, Spain, 4–8 May 2020; IEEE: Piscataway, NJ, USA, 2020; pp. 4866–4870.

6. Luo, R.; Yan, L.; Deng, P.; Kuang, Y. Hybrid TOA/AOA Virtual Station Localization Based on Scattering Signal Identification for GNSS-Denied Urban or Indoor NLOS Environments. *Appl. Sci.* **2022**, *12*, 12157. [CrossRef]
7. Li, J.; Conan, J.; Pierre, S. Position location of mobile terminal in wireless MIMO communication systems. *J. Commun. Netw.* **2007**, *9*, 254–264. [CrossRef]
8. Fellhauer, F.; Lassen, J.; Jaber, A.; Loghin, N.; Ten Brink, S. Non-line-of-sight positioning for mmWave communications. In Proceedings of the 2018 IEEE 19th International Workshop on Signal Processing Advances in Wireless Communications (SPAWC), Kalamata, Greece, 25–28 June 2018; IEEE: Piscataway, NJ, USA, 2018; pp. 1–5.
9. Shastri, A.; Valecha, N.; Bashirov, E.; Tataria, H.; Lentmaier, M.; Tufvesson, F.; Rossi, M.; Casari, P. A review of millimeter wave device-based localization and device-free sensing technologies and applications. *IEEE Commun. Surv. Tutor.* **2022**, *24*, 1708–1749. [CrossRef]
10. Perdana, D.; Indra Tanaya, I.M.A.; Marwan, A.A.; Akhyar, F. Evaluation of a High-Accuracy Indoor-Positioning System with Wi-Fi Time of Flight (ToF) and Deep Learning. *J. Comput. Netw. Commun.* **2023**, *2023*, 6777759. [CrossRef]
11. Shastri, A.; Palacios, J.; Casari, P. Millimeter wave localization with imperfect training data using shallow neural networks. In Proceedings of the 2022 IEEE Wireless Communications and Networking Conference (WCNC), Austin, TX, USA, 10–13 April 2022; IEEE: Piscataway, NJ, USA, 2022; pp. 674–679.
12. Yu, S.; Zhu, J.; Lv, C. A Quantum Annealing Bat Algorithm for Node Localization in Wireless Sensor Networks. *Sensors* **2023**, *23*, 782. [CrossRef] [PubMed]
13. Luo, Q.; Yang, K.; Yan, X.; Li, J.; Wang, C.; Zhou, Z. An Improved Trilateration Positioning Algorithm with Anchor Node Combination and K-Means Clustering. *Sensors* **2022**, *22*, 6085. [CrossRef] [PubMed]
14. Fortes, J.; Švingál, M.; Porteleky, T.; Jurík, P.; Drutarovský, M. Positioning and Tracking of Multiple Humans Moving in Small Rooms Based on a One-Transmitter–Two-Receiver UWB Radar Configuration. *Sensors* **2022**, *22*, 5228. [CrossRef] [PubMed]
15. Kolakowski, M. Improving accuracy and reliability of bluetooth low-Energy-Based localization systems using proximity sensors. *Appl. Sci.* **2019**, *9*, 4081. [CrossRef]
16. Oh, S.H.; Kim, J.G. WiFi positioning in 3GPP indoor office with modified particle swarm optimization. *Appl. Sci.* **2021**, *11*, 9522. [CrossRef]
17. Ma, Y.; Selby, N.; Adib, F. Minding the billions: Ultra-wideband localization for deployed RFID tags. In Proceedings of the 23rd Annual International Conference on Mobile Computing and Networking, Snowbird, UT, USA, 16–20 October 2017; pp. 248–260.
18. Shen, M.; Wang, Y.; Jiang, Y.; Ji, H.; Wang, B.; Huang, Z. A New Positioning Method Based on Multiple Ultrasonic Sensors for Autonomous Mobile Robot. *Sensors* **2020**, *20*, 17. [CrossRef] [PubMed]
19. Derkach, M.; Matiuk, D.; Skarga-Bandurova, I. Obstacle Avoidance Algorithm for Small Autonomous Mobile Robot Equipped with Ultrasonic Sensors. In Proceedings of the 2020 IEEE 11th International Conference on Dependable Systems, Services and Technologies (DESSERT), Kyiv, Ukraine, 14–18 May 2020; pp. 236–241. [CrossRef]
20. Saad, M.M.; Bleakley, C.J.; Ballal, T.; Dobson, S. High-Accuracy Reference-Free Ultrasonic Location Estimation. *IEEE Trans. Instrum. Meas.* **2012**, *61*, 1561–1570. [CrossRef]
21. Nagy, C.; Biró-Ambrus, Z.; Márton, L. Ultrasound-Based Indoor Robot Localization Using Ambient Temperature Compensation. *Acta Univ. Sapientiae Electr. Mech. Eng.* **2016**, *8*, 19–28. [CrossRef]

applied sciences

MDPI

Article

Kinematic Parameter Identification for a Parallel Robot with an Improved Particle Swarm Optimization Algorithm

Dayong Yu

School of Mechanical Engineering, University of Shanghai for Science and Technology, Shanghai 200093, China; yudayong@usst.edu.cn

Abstract: The spacecraft docking motion simulation system for on-orbit docking plays a very important role in some theoretical research and engineering application fields. The parallel robot utilized in the spacecraft docking simulation system requires high positioning and orientation accuracy to achieve better simulation results. A novel kinematic parameter identification method with an improved particle swarm optimization (PSO) algorithm is proposed to enhance positioning and orientation accuracy of the parallel robot. A fitness function is established using these residuals between the measured and computed poses by a coordinate measuring machine and forward kinematics. The kinematic parameter identification problem is turned into a high-dimensional nonlinear optimization in which the unknown kinematic parameter errors are regarded as optimal variables. The optimal variables are solved by the proposed improved PSO algorithm. The mean values of the positioning and orientation errors are reduced from 4.3268 mm and 0.2221 deg to 0.7692 mm and 0.0674 deg, respectively. The proposed kinematic parameter identification method increases the positioning accuracy mean by 22.26% and the orientation accuracy mean by 32.80% compared with the least squares method. The kinematic parameter identification method with the improved PSO algorithm can effectively enhance positioning and orientation accuracy of the parallel robot for docking motion simulation.

Keywords: parallel robot; parameter identification; particle swarm optimization algorithm; pose accuracy; nonlinear least squares

1. Introduction

Parallel robots have obvious advantages in structural stiffness, carrying capacity, and dynamic characteristics compared with series robots. Parallel robots have been widely used in these fields, such as motion simulation [1–5], mechanical manufacturing [6–10], and surgery operating [11–15].

In these application fields, there are different requirements for the positioning and orientation accuracy performance of parallel robots. Recently, the requirement for positioning and orientation accuracy performance of parallel robots is becoming increasingly stringent and detailed [16–18]. The actual kinematic parameters and the nominal kinematic parameters of parallel robots are usually inconsistent due to manufacturing and install errors, which will result in the positioning and orientation accuracy reduction of parallel robots in practical applications. At present, kinematic calibration is the most feasible and effective technique to improve the positioning and orientation accuracy of parallel robots. The technique usually consists of four steps, including kinematic parameter error modeling, terminal pose measurement, kinematic parameter error identification, and kinematic parameter error compensation. Firstly, the kinematic parameter error model was established under appropriate modeling criteria. Secondly, the terminal position and orientation were precisely measured by using available external measuring instruments. Subsequently, a specific parameter identification method was needed to determine the kinematic parameter errors. Finally, the nominal kinematic parameters residing in the local controller were updated to guarantee high positioning and orientation accuracy of parallel robots.

The kinematic parameter error identification is the most important step of kinematic calibration. The kinematic parameter error identifications are usually expressed as a standard linear or nonlinear least squares optimization [19–22]. The optimization problem can be solved by conventional or intelligent algorithms. Verner et al. [23] presented a new optimal kinematic calibration method to obtain the geometric errors by utilizing the least error-sensitive areas in the workspace of parallel kinematic machines. Ni et al. [24] established an error-mapping model and proposed parameter identification of a novel parallel manipulator with full-circle rotation with double ball–bar measurement data. Liu et al. [25] analyzed the Newton iterative parameter identification method and developed a modified parameter identification method of a 3-DOF parallel tool head to simplify the algorithm parameter selection in the regularization identification method. Fu et al. [26] constructed a dimensionless error model considering the geometrical characteristics of a Stewart platform and investigated the kinematic parameter error identification accuracy of the Stewart platform based on the least squares algorithm. He et al. [27] investigated the identifiability of kinematic parameter errors of a 6-RUHU parallel manipulator and obtained a minimal identification model to realize kinematic calibration with a smaller number of random measurements. The kinematic parameter errors were identified by using the linear least squares algorithm. However, the least squares algorithm is liable to fall into local optimal solutions.

A kinematic calibration of a portable coordinate measuring machine with parallel-joint based on the Gauss–Newton algorithm was put forward by Li et al. [28]. Mao et al. [29] converted the kinematic calibration problem of serial robots into a separable nonlinear least squares problem. Then, the optimal kinematic parameters were identified by using the Gauss–Newton iterative algorithm. A comprehensive kinematic calibration procedure of hexapod machine tools was proposed by Nategh and Agheli [30], and the actual kinematic parameters were solved by minimizing a cost function based on the Levenberg–Marquardt algorithm. Majarena et al. [31] analyzed different cost functions for the kinematic parameter identification of parallel robots, and kinematic parameters of these cost functions were identified by using the Levenberg–Marquardt algorithm. Luo et al. [32] set up an actual kinematic model of a 4PPa-2PaR parallel robot based on vector theory, and the kinematic parameter errors were identified based on the Levenberg–Marquardt algorithm. In general, the nonlinear least squares algorithm can only obtain the local optimal solutions, which is not necessarily for the global optimal solutions.

Nguyen et al. [33] proposed an integration kinematic calibration method, and the robot geometric parameter errors were identified by using an extended Kalman filtering algorithm. Qi et al. [34] also proposed an integration kinematic calibration method combining quantitative errors with random errors. Then, the quantitative errors were identified by using an extended Kalman filtering algorithm. A kinematic calibration method of a novel 5-DOF hybrid robot was presented by Yin et al. [35], and a reliable and stable geometric error identification strategy was obtained based on an extended Kalman filter algorithm. A kinematic calibration method of space manipulators using an extended Kalman filter was put forward by Wang et al. [36], which can overcome the adverse influence of measuring noise and reduce the identification time. Yang et al. [37] constructed a kinematic calibration model combining an extended Kalman filter with residual neural networks, and geometric parameter errors were addressed by using an extended Kalman filter algorithm. A kinematic parameter error identification method for a GOOGOL GRB3016 based on an extended Kalman filter algorithm was presented by Du and Zhang [38]. A new kinematic parameter error identification method through particle filter incorporated extended Kalman filter algorithm was presented by Jiang et al. [39], which can deal with non-Gauss noise and a high nonlinear kinematics. An online kinematic parameter identification method was proposed by Du et al. [40] to correct kinematic parameters without stopping the robot, and the kinematic parameters were reckoned by using an unscented Kalman filter. Gao et al. [41] presented a novel hybrid kinematic parameter identification method integrating an extended Kalman filter and regularized particle filter to calibrate an industrial robot. Compared with the least squares algorithm, the extended Kalman filtering algorithm is robust and has good compensation effect in the verification configurations.

Feng et al. [42] applied three intelligent optimization algorithms to calibrate a planar 2-DOF parallel manipulator. A kinematic calibration of a 2-DOF parallel mechanism was addressed by Sun et al. [43], and the kinematic parameters were directly solved by using a hybrid genetic algorithm. Fei et al. [44] presented an improved genetic algorithm to calibrate a five-axis measuring machine. A novel kinematic calibration method of a parallel mechanism was proposed by Fan et al. [45], and the model parameters of the direct kinematics were identified by using a genetic algorithm. Huang et al. [46] constructed a normalized error model including geometric and non-geometric errors of a 3-TT parallel mechanism to enhance the kinematic calibration effect, and the nonredundant kinematic parameters were identified by using a genetic algorithm. Gao et al. [47] proposed a novel kinematic parameter identification method for industrial robots via BP neural network-incorporated PSO algorithm, and this method can reduce iterations and enhance convergence rate compared with the standard PSO algorithm. An improved PSO algorithm was proposed by Wang et al. [48] to choose the optimal pose measurement set in kinematic calibration, and the algorithm can realize the optimal pose set selection in a measurable workspace. A kinematic parameter identification method based on an improved PSO algorithm was put forward by Li et al. [49], and this method can reduce running time and enhance solving efficiency by linearly decreasing the inertia factor and choosing the particle number with roulette wheel selection. An improved beetle swarm optimization algorithm was put forward by Chen et al. [50] to calibrate an industrial robot, and this algorithm can avoid local optimal solutions and enhance searching reliability. Zhong et al. [51] proposed an improved whale swarm algorithm to calibrate a two-segment hydraulic leg, and the kinematic calibration method can improve the tracking accuracy. Gong et al. [52] proposed a hybrid particle swarm–artificial bee colony optimization algorithm to identify structural model parameters, and this algorithm has stronger searching ability and higher optimizing accuracy. Fang et al. [53] proposed a kinematic calibration for serial robots and parallel robots, and kinematic parameter errors were determined by using the quantum-behaved PSO algorithm. In some respects, the PSO algorithm is similar to the genetic algorithm, but the PSO algorithm requires no extra operations, such as crossover, and fewer adjustment parameters.

The principle of the standard PSO algorithm can be summarized as finding the optimal solution through collaboration and information sharing among the group. The superiorities of the standard PSO algorithm lie in simple modeling, convenient implementation, rapid convergence, and achieving global optimal solutions. The PSO algorithm can overcome shortcomings of conventional kinematic parameter error identification methods in kinematic calibration, which need to depend on specific error models and complex solving procedures. Therefore, the PSO algorithm is suitable for kinematic parameter error identification of parallel robots in kinematic calibration.

In this article, a kinematic parameter identification method based on forward kinematics is presented for a parallel robot in a docking motion simulation system. The kinematic parameter identification is represented as an optimal problem for a discrete objective function. The global Newton–Raphson algorithm is used to solve the forward kinematics of the parallel robot. Since there is no analytical solution to the forward kinematics, the PSO algorithm is suitable to solve the discrete objective function. In order to solve this problem, improving the PSO algorithm is necessary. The proposed kinematic parameter identification method is applied to determine the kinematic parameter errors of the parallel robot.

The rest of this paper is organized as follows: The kinematic models of the parallel robot are established in Section 2. The identification model and kinematic parameter identification method with an improved particle swarm optimization algorithm are presented in Section 3. The identification simulations and calibration experiments are provided in Section 4. Some conclusions are given in Section 5.

2. Kinematic Modeling

2.1. Coordinate Frame

A prototype docking motion simulation system was built at the Harbin Institute of Technology to simulate on-orbit docking motion of the spacecrafts and test the comprehensive performance of the docking mechanisms. The parallel robot consists of a fixed base, a mobile platform, six identical length changeable hydraulic cylinders, and twelve Hooke's joints. The parallel robot has six degrees of freedom.

The coordinate frame is established and shown in Figure 1 to analyze the kinematic performance of the parallel robot. The static coordinate frame O_B-xyz is set up at the hexagon center of the fixed base. The dynamic coordinate frame O_D-xyz is set up at the hexagon center of the mobile platform. The motion output of the parallel robot is expressed by the pose of the mobile platform, denoted by position vector $\boldsymbol{t}$ ($\boldsymbol{t}$ = [x y z]) and orientation vector $\boldsymbol{\omega}$ ($\boldsymbol{\omega}$ = [α β γ]).

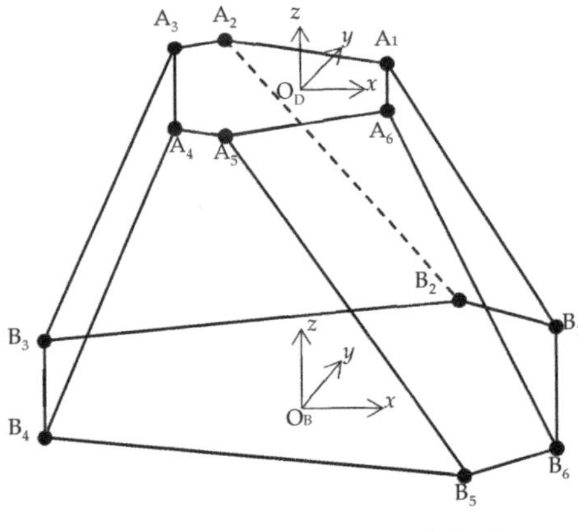

Figure 1. Coordinate frame of the parallel robot.

2.2. Kinematics Model

The rotation transformation matrix from O_D-xyz to O_B-xyz is $\boldsymbol{R}$, which is denoted by the following expression:

$$\boldsymbol{R} = \begin{bmatrix} cos\beta cos\gamma & sin\alpha sin\beta cos\gamma - cos\alpha sin\gamma & cos\alpha sin\beta cos\gamma + sin\alpha sin\gamma \\ cos\beta sin\gamma & sin\alpha sin\beta sin\gamma + cos\alpha cos\gamma & cos\alpha sin\beta sin\gamma - sin\alpha cos\gamma \\ -sin\beta & sin\alpha cos\beta & cos\alpha cos\beta \end{bmatrix} \quad (1)$$

The positions of the six Hooke's joints connected to the mobile platform of the parallel robot are denoted by $\boldsymbol{a}_i$ ($\boldsymbol{a}_i$ = [a_{ix} a_{iy} a_{iz}], i = 1, 2, ..., 6) in the dynamic coordinate frame O_D-xyz. The positions of the six Hooke's joints connected to the fix base are denoted by $\boldsymbol{b}_i$ ($\boldsymbol{b}_i$ = [b_{ix} b_{iy} b_{iz}], i = 1, 2, ..., 6) in the static coordinate frame O_B-xyz. Thus, the hydraulic cylinder length l_i of the parallel robot can be expressed as follows:

$$l_i = \left\| \boldsymbol{R}\boldsymbol{a}_i^T + \boldsymbol{t}^T - \boldsymbol{b}_i^T \right\| \quad i = 1,\ 2,\ \ldots,\ 6 \quad (2)$$

The inverse kinematics of the parallel robot can be denoted by the following expression:

$$\boldsymbol{l} = f_I(\boldsymbol{p}, \boldsymbol{a}, \boldsymbol{b}) \quad (3)$$

where $\boldsymbol{p} = [\boldsymbol{t}\ \boldsymbol{\omega}]^T$, $\boldsymbol{a} = [\boldsymbol{a}_1\ \boldsymbol{a}_2\ \boldsymbol{a}_3\ \boldsymbol{a}_4\ \boldsymbol{a}_5\ \boldsymbol{a}_6]^T$, $\boldsymbol{b} = [\boldsymbol{b}_1\ \boldsymbol{b}_2\ \boldsymbol{b}_3\ \boldsymbol{b}_4\ \boldsymbol{b}_5\ \boldsymbol{b}_6]^T$, $\boldsymbol{l} = [l_1\ l_2\ l_3\ l_4\ l_5\ l_6]^T$. The hydraulic cylinder length vector $\boldsymbol{l}$ can be calculated by Equation (3) under the given pose of the parallel robot. It is obvious that the solution of the inverse kinematics of the parallel robot is analytical. The calculated stroke length vector $\boldsymbol{l}_c$ of the hydraulic cylinder is expressed as follows:

$$\boldsymbol{l}_c = \boldsymbol{l} - \boldsymbol{l}_0 \tag{4}$$

where $\boldsymbol{l}_0$ ($\boldsymbol{l}_0$ = $[l_{01}\ l_{02}\ l_{03}\ l_{04}\ l_{05}\ l_{06}]^{\mathrm{T}}$) is the initial length vector of the hydraulic cylinder at the initial pose of the parallel robot $\boldsymbol{p}_0$ ($\boldsymbol{p}_0$ = $[0\ 0\ 0\ 0\ 0\ 0]^{\mathrm{T}}$).

The forward kinematics of the parallel robot can be denoted by the following expression:

$$\boldsymbol{p} = f_F(\boldsymbol{l}, \boldsymbol{a}, \boldsymbol{b}) \tag{5}$$

The forward kinematics is composed of six nonlinear equations. The pose of the mobile platform can be calculated by substituting the given hydraulic cylinder length into Equation (5). Although inverse kinematics can offer an analytical solution, forward kinematics is a complex and highly nonlinear process. The global Newton–Raphson algorithm [54] is used to calculate the pose of the mobile platform. This algorithm can obtain a set of high-precision numerical convergent solutions with any initial guesses in the workspace of the parallel robot. Equation (5) can be expanded into the standard form for the global Newton–Raphson algorithm and expressed as follows:

$$\begin{cases} f_1 = \left\| \boldsymbol{R}\boldsymbol{a}_1^T + \boldsymbol{t}^T - \boldsymbol{b}_1^T \right\| - l_{01} - l_{m1} = 0 \\ f_2 = \left\| \boldsymbol{R}\boldsymbol{a}_2^T + \boldsymbol{t}^T - \boldsymbol{b}_2^T \right\| - l_{02} - l_{m2} = 0 \\ f_3 = \left\| \boldsymbol{R}\boldsymbol{a}_3^T + \boldsymbol{t}^T - \boldsymbol{b}_3^T \right\| - l_{03} - l_{m3} = 0 \\ f_4 = \left\| \boldsymbol{R}\boldsymbol{a}_4^T + \boldsymbol{t}^T - \boldsymbol{b}_4^T \right\| - l_{04} - l_{m4} = 0 \\ f_5 = \left\| \boldsymbol{R}\boldsymbol{a}_5^T + \boldsymbol{t}^T - \boldsymbol{b}_5^T \right\| - l_{05} - l_{m5} = 0 \\ f_6 = \left\| \boldsymbol{R}\boldsymbol{a}_6^T + \boldsymbol{t}^T - \boldsymbol{b}_6^T \right\| - l_{06} - l_{m6} = 0 \end{cases} \tag{6}$$

where $\boldsymbol{l}_{\mathrm{m}}$ ($\boldsymbol{l}_{\mathrm{m}}$ = $[l_{\mathrm{m}1}\ l_{\mathrm{m}2}\ l_{\mathrm{m}3}\ l_{\mathrm{m}4}\ l_{\mathrm{m}5}\ l_{\mathrm{m}6}]^{\mathrm{T}}$) is the measured stroke length vector of the hydraulic cylinder.

The Jacobi matrix for solving Equation (6) is expressed as follows:

$$\boldsymbol{J} = \begin{bmatrix} \frac{\partial f_1}{\partial x} & \frac{\partial f_1}{\partial y} & \frac{\partial f_1}{\partial z} & \frac{\partial f_1}{\partial \alpha} & \frac{\partial f_1}{\partial \beta} & \frac{\partial f_1}{\partial \gamma} \\ \frac{\partial f_2}{\partial x} & \frac{\partial f_2}{\partial y} & \frac{\partial f_2}{\partial z} & \frac{\partial f_2}{\partial \alpha} & \frac{\partial f_2}{\partial \beta} & \frac{\partial f_2}{\partial \gamma} \\ \frac{\partial f_3}{\partial x} & \frac{\partial f_3}{\partial y} & \frac{\partial f_3}{\partial z} & \frac{\partial f_3}{\partial \alpha} & \frac{\partial f_3}{\partial \beta} & \frac{\partial f_3}{\partial \gamma} \\ \frac{\partial f_4}{\partial x} & \frac{\partial f_4}{\partial y} & \frac{\partial f_4}{\partial z} & \frac{\partial f_4}{\partial \alpha} & \frac{\partial f_4}{\partial \beta} & \frac{\partial f_4}{\partial \gamma} \\ \frac{\partial f_5}{\partial x} & \frac{\partial f_5}{\partial y} & \frac{\partial f_5}{\partial z} & \frac{\partial f_5}{\partial \alpha} & \frac{\partial f_5}{\partial \beta} & \frac{\partial f_5}{\partial \gamma} \\ \frac{\partial f_6}{\partial x} & \frac{\partial f_6}{\partial y} & \frac{\partial f_6}{\partial z} & \frac{\partial f_6}{\partial \alpha} & \frac{\partial f_6}{\partial \beta} & \frac{\partial f_6}{\partial \gamma} \end{bmatrix} \tag{7}$$

The iterative modification process can be expressed as follows:

$$\begin{cases} \boldsymbol{J}\delta\boldsymbol{p}^{(k)} = -\lambda^{(k)} \cdot \boldsymbol{f}\left(\boldsymbol{p}^{(k)}\right) \\ \boldsymbol{p}^{(k+1)} = \boldsymbol{p}^{(k)} + \delta\boldsymbol{p}^{(k)} \end{cases} \quad 0 < \lambda^{(k)} < 1 \tag{8}$$

where $\boldsymbol{p}^{(k)}$ is the kth step iterative value, $\boldsymbol{f}$ ($\boldsymbol{f}$ = $[f_1\ f_2\ f_3\ f_4\ f_5\ f_6]^{\mathrm{T}}$) is the vector function describing the difference between the calculated stroke lengths with Equation (4) and the measured stroke lengths with displacement sensors, and $\lambda^{(k)}$ is the monotonic descent operator to ensure the convergence of iteration, which can be automatically selected and satisfies Equations (9) and (10).

$$\left\|\boldsymbol{f}\left(\boldsymbol{p}^{(k-1)}\right)\right\| < \left\|\boldsymbol{f}\left(\boldsymbol{p}^{(k)}\right)\right\| \tag{9}$$

$$\boldsymbol{p} \in \mathrm{D} \tag{10}$$

where D is the workspace without singularity of the parallel robot.

The iterative termination condition of the global Newton–Raphson algorithm is expressed as follows:

$$\left| max\left\{ f\left(p^{(k)}\right) - f\left(p^{(k-1)}\right) \right\} \right| < \varepsilon \quad k \leq N \tag{11}$$

where ε is defined as the permissible error (required precision level of the final solution) of the proximity of the calculated values of the vector function f between adjacent iterations, and it is an infinite decimal. N is defined as the maximal number of iterations in a control cycle.

2.3. Kinematic Model with Errors

The actual and nominal forward kinematics of the parallel robot are inconsistent due to the manufacturing and installation errors. The pose of the parallel robot considering the manufacturing and installation errors can be denoted by position vector $\boldsymbol{t}'$ ($\boldsymbol{t}'$ = [x' y' z']) and orientation vector $\boldsymbol{\omega}'$ ($\boldsymbol{\omega}'$ = [α' β' γ']). Thus, the rotation transformation matrix of the mobile platform relative to the fixed base can be expressed as follows:

$$\boldsymbol{R}' = \begin{bmatrix} cos\beta' cos\gamma' & sin\alpha' sin\beta' cos\gamma' - cos\alpha' sin\gamma' & cos\alpha' sin\beta' cos\gamma' + sin\alpha' sin\gamma' \\ cos\beta' sin\gamma' & sin\alpha' sin\beta' sin\gamma' + cos\alpha' cos\gamma' & cos\alpha' sin\beta' sin\gamma' - sin\alpha' cos\gamma' \\ -sin\beta' & sin\alpha' cos\beta' & cos\alpha' cos\beta' \end{bmatrix} \tag{12}$$

Therefore, an actual forward kinematics including the manufacturing and installation errors is constructed. By substituting Equation (12) and position vector $\boldsymbol{t}'$ into Equation (6), the actual forward kinematics of the parallel robot can be expressed as follows:

$$\begin{cases} f_1 = \left\| \boldsymbol{R}'\boldsymbol{a}'^T_1 + \boldsymbol{t}'^T - \boldsymbol{b}'^T_1 \right\| - l'_{01} - l_{m1} = 0 \\ f_2 = \left\| \boldsymbol{R}'\boldsymbol{a}'^T_2 + \boldsymbol{t}'^T - \boldsymbol{b}'^T_2 \right\| - l'_{02} - l_{m2} = 0 \\ f_3 = \left\| \boldsymbol{R}'\boldsymbol{a}'^T_3 + \boldsymbol{t}'^T - \boldsymbol{b}'^T_3 \right\| - l'_{03} - l_{m3} = 0 \\ f_4 = \left\| \boldsymbol{R}'\boldsymbol{a}'^T_4 + \boldsymbol{t}'^T - \boldsymbol{b}'^T_4 \right\| - l'_{04} - l_{m4} = 0 \\ f_5 = \left\| \boldsymbol{R}'\boldsymbol{a}'^T_5 + \boldsymbol{t}'^T - \boldsymbol{b}'^T_5 \right\| - l'_{05} - l_{m5} = 0 \\ f_6 = \left\| \boldsymbol{R}'\boldsymbol{a}'^T_6 + \boldsymbol{t}'^T - \boldsymbol{b}'^T_6 \right\| - l'_{06} - l_{m6} = 0 \end{cases} \tag{13}$$

The global Newton–Raphson algorithm, as introduced in Section 2.2, can also be utilized to solve the actual forward kinematics of the parallel robot. Accordingly, the Jacobi matrix to solve the actual forward kinematics is expressed as follows:

$$\boldsymbol{J} = \begin{bmatrix} \frac{\partial f_1}{\partial x'} & \frac{\partial f_1}{\partial y'} & \frac{\partial f_1}{\partial z'} & \frac{\partial f_1}{\partial \alpha'} & \frac{\partial f_1}{\partial \beta'} & \frac{\partial f_1}{\partial \gamma'} \\ \frac{\partial f_2}{\partial x'} & \frac{\partial f_2}{\partial y'} & \frac{\partial f_2}{\partial z'} & \frac{\partial f_2}{\partial \alpha'} & \frac{\partial f_2}{\partial \beta'} & \frac{\partial f_2}{\partial \gamma'} \\ \frac{\partial f_3}{\partial x'} & \frac{\partial f_3}{\partial y'} & \frac{\partial f_3}{\partial z'} & \frac{\partial f_3}{\partial \alpha'} & \frac{\partial f_3}{\partial \beta'} & \frac{\partial f_3}{\partial \gamma'} \\ \frac{\partial f_4}{\partial x'} & \frac{\partial f_4}{\partial y'} & \frac{\partial f_4}{\partial z'} & \frac{\partial f_4}{\partial \alpha'} & \frac{\partial f_4}{\partial \beta'} & \frac{\partial f_4}{\partial \gamma'} \\ \frac{\partial f_5}{\partial x'} & \frac{\partial f_5}{\partial y'} & \frac{\partial f_5}{\partial z'} & \frac{\partial f_5}{\partial \alpha'} & \frac{\partial f_5}{\partial \beta'} & \frac{\partial f_5}{\partial \gamma'} \\ \frac{\partial f_6}{\partial x'} & \frac{\partial f_6}{\partial y'} & \frac{\partial f_6}{\partial z'} & \frac{\partial f_6}{\partial \alpha'} & \frac{\partial f_6}{\partial \beta'} & \frac{\partial f_6}{\partial \gamma'} \end{bmatrix} \tag{14}$$

The iterative modification process is described as follows:

$$\begin{cases} \boldsymbol{J}\delta\boldsymbol{p}' = -\lambda \cdot \boldsymbol{f}(\boldsymbol{p}') \\ \boldsymbol{p}'^{(k+1)} = \boldsymbol{p}'^{(k)} + \delta\boldsymbol{p}'^{(k)} \end{cases} \quad 0 < \lambda < 1 \tag{15}$$

where $\boldsymbol{p}'$ = [$\boldsymbol{t}'$ $\boldsymbol{\omega}'$] ^{T} and $\boldsymbol{p}'^{(k)}$ is the kth step iterative value.

3. Kinematic Parameter Identification Method

3.1. Identification Model

In this paper, the actual kinematic parameters of the parallel robot can be identified by utilizing the measured pose data. These actual kinematic parameters include the position coordinates of the twelve Hooke's joints and the initial lengths of the six hydraulic cylinders. To facilitate the following kinematic parameter identification modeling, the nominal solution of the inverse kinematics of the parallel robot can be denoted by $\boldsymbol{l}_{ci}$ and expressed as follows:

$$\boldsymbol{l}_{ci} = f_I(\boldsymbol{p}_{ci}, \boldsymbol{\rho}_0) \tag{16}$$

where $i = 1, 2, \ldots, n$ is the selected pose number in the identification process, $\boldsymbol{l}_{ci}$ is the stroke length vector of the hydraulic cylinder, $\boldsymbol{p}_{ci}$ is the nominal pose vector of the mobile platform, and $\boldsymbol{\rho}_0$ is the nominal kinematic parameter vector of the parallel robot.

Similarly, the nominal solution of the forward kinematics of the parallel robot can be denoted by $\boldsymbol{p}_{ci}$ and simplified as follows:

$$\boldsymbol{p}_{ci} = f_F(\boldsymbol{l}_{ci}, \boldsymbol{\rho}_0) \tag{17}$$

When the kinematic parameter errors produced by the manufacturing and installation processes are taken into account, the actual kinematic parameter vector of the parallel robot is denoted by $\boldsymbol{\rho}$. Hence, the calculated pose of the parallel robot is expressed as follows:

$$\boldsymbol{p}_i = f_F(f_I(\boldsymbol{p}_{ci}, \boldsymbol{\rho}_0), \boldsymbol{\rho}) \tag{18}$$

The pose error of the parallel robot can be given by the following expression:

$$\boldsymbol{e}_i = \boldsymbol{p}_{mi} - \boldsymbol{p}_i \tag{19}$$

where $\boldsymbol{p}_{\mathrm{mi}}$ is the vector of the measured pose with a coordinate measuring machine.

Measurement error is inevitable in the actual pose measuring experiment. Let $\boldsymbol{p}_{\mathrm{i}}'$ be the nominal pose vector and $\boldsymbol{\varepsilon}_i$ be the measurement error vector for the ith measured pose. Thus, the vector of the measured pose can be expressed as follows:

$$\boldsymbol{p}_{mi} = \boldsymbol{p}_i' + \boldsymbol{\varepsilon}_i \tag{20}$$

By substituting Equations (18) and (20) into Equation (19), the pose error of the parallel robot can be expressed as follows:

$$\boldsymbol{e}_i = (\boldsymbol{p}_i' + \boldsymbol{\varepsilon}_i) - f_F(f_I(\boldsymbol{p}_{ci}, \boldsymbol{\rho}_0), \boldsymbol{\rho}) \tag{21}$$

It is well known that the measurement error vector $\boldsymbol{\varepsilon}_i$ is uncertain. In addition, $\boldsymbol{\varepsilon}_i$ is changeable in each measuring process. For n = 7, although an approximate solution of the actual kinematic parameters of the parallel robot can be calculated, the solution only decreases local pose error and is sensitive to the measurement error. In order to improve the global pose accuracy of the parallel robot and reduce the sensitivity of the pose measurement error, the poses of more than 7 should be measured in the kinematic parameter identification process. Thus, the optimization method can be used to identify the actual kinematic parameters of the parallel robot. Let L be the optimization objective function, and take the minimum of the sum of squares of $\boldsymbol{e}_i$ as the objective function. The objective function L can be expressed as follows:

$$L = minE = min\left\{\sum_{i=1}^{n} \boldsymbol{e}_i^T \boldsymbol{e}_i\right\} \tag{22}$$

3.2. Parameter Identification with an Improved PSO Algorithm

After the objective function is determined, another important problem is to select a proper kinematic parameter identification algorithm to minimize the objective function. Most of the conventional kinematic parameter identification algorithms in kinematic calibration are based on the derivative of the objective function. As shown in Equations (13)–(18), for the given actual kinematic parameter vector $\boldsymbol{\rho}$, the pose $\boldsymbol{p}_i$ at the ith identification point can be computed by the global Newton–Raphson algorithm with the forward kinematics. Therefore, the objective function L cannot be expressed as an analytic formula of the actual kinematic parameter vector $\boldsymbol{\rho}$. Furthermore, the derivative of the objective function L cannot also be analytically expressed.

The PSO algorithm, which is proposed by Kennedy and Eberhart [55], can be used to solve the optimization problem with multiple variables. The basic idea of the PSO algorithm is to find the optimal solution by cooperation and information sharing among individuals in the group. The PSO algorithm has no restrictions on the derivation and continuity of the objective function, which is suitable for the discrete optimization problem. It is well known that the PSO algorithm has few parameters to be adjusted and a simple iterative process. Hence, the PSO algorithm can be directly used to identify the actual kinematic parameters of the parallel robot. The identification process based on the PSO algorithm is described as follows:

The dimensionality space of the actual kinematic parameter Identification of the parallel robot is represented by d (d = 42), and the population space $\boldsymbol{X}$ ($\boldsymbol{X} = [\boldsymbol{X}_1\ \boldsymbol{X}_2 \ldots \boldsymbol{X}_N]$) includes N random particles. A feasible solution to the actual kinematic parameter identification of the parallel robot can be expressed as $\boldsymbol{X}_i$ ($\boldsymbol{X}_i = [x_{i1}\ x_{i2} \ldots x_{id}]$), which describes the position of particle I in the search space. The fitness value of particle I can be computed by using Equation (22) and expressed as $E(\boldsymbol{X}_i)$. The position update of particle I can be performed by using Equation (23).

$$\boldsymbol{X}_i^{t+1} = \boldsymbol{X}_i^t + \boldsymbol{V}_i^{t+1} \tag{23}$$

where t (t = 1, 2, . . ., G) is the iterative number, G is the maximum iterative number, and $\boldsymbol{V}_i$ ($\boldsymbol{V}_{\mathrm{i}} = [v_{\mathrm{i}1}\ v_{i2} \ldots v_{i\mathrm{d}}]$) represents the velocity of particle i. The velocity update of particle I can be performed by using Equation (24).

$$\boldsymbol{V}_i^{t+1} = w^t\boldsymbol{V}_i^t + c_1r_1(\boldsymbol{P}_i^t - \boldsymbol{X}_i^t) + c_2r_2\left(\boldsymbol{P}_g^t - \boldsymbol{X}_i^t\right) \tag{24}$$

where w is the inertia weight which represents the inheritance of the current velocity, c_1 is the local learning factor, c_2 is the global learning factor, r_1 and r_2 are random numbers between 0 and 1, $\boldsymbol{P}_i$ ($\boldsymbol{P}_i = [p_{i1}\ p_{i2} \ldots p_{id}]$) is the local optimal position vector, and $\boldsymbol{P}_g$ ($\boldsymbol{P}_g = [p_{g1}\ p_{g2} \ldots p_{gd}]$) is the global optimal position vector.

The actual kinematic parameter identification process based on the PSO algorithm begins with the initialization of a set of random particles, after which the fitness value of each particle is calculated so that the best of these fitness values is recorded as the global optimum. Thereafter, the position and velocity of each particle in each iteration are modified based on Equations (23) and (24). Next, the new fitness value of each particle is compared with the current global optimum fitness value, and the global optimal position vector is updated. The iterative process is repeated until meeting the iterative termination conditions and the global optimal position vector is obtained.

In general, the PSO algorithm is used to solve unconstrained optimization problems. Therefore, the PSO algorithm lacks the mechanism to handle the boundary constraints. In addition, the actual kinematic parameters of the parallel robot that need to be identified are up to forty-two, so the convergence speed and identification precision need to be improved while guaranteeing the excellent global search capability. Several modification strategies were adopted and applied to the PSO algorithm to obtain the excellent actual kinematic parameter identification result.

The boundary processing strategy of the PSO algorithm is to adjust the velocity direction of the out-of-boundary particles to the opposite direction in order to prevent the search from entering a stagnation state and falling into a local optimal trap. A new boundary processing strategy [50] is adopted and expressed as follows:

$$\mathbf{X}_i^{t+1} = \begin{cases} \mathbf{X}_i^t - \lambda \mathbf{V}_i^{t+1} & \mathbf{X}_i^{t+1} \geq \mathbf{X}_{max} \ or \ \mathbf{X}_i^{t+1} \leq \mathbf{X}_{min} \\ \mathbf{X}_i^t + \lambda \mathbf{V}_i^{t+1} & \mathbf{X}_{min} \leq \mathbf{X}_i^{t+1} \leq \mathbf{X}_{max} \end{cases} \tag{25}$$

where $\mathbf{X}_{max}$ and $\mathbf{X}_{min}$ are the upper and lower boundary of the search space, respectively.

For the sake of strengthening the local and global optimal search ability and enhancing convergence speed, a nonlinear inertia weight adjustment strategy [56] is adopted and expressed as follows:

$$w = \begin{cases} w_{max} - (w_{max} - w_{min})\frac{E(\mathbf{X}_i) - E_{min}}{E_{ave} - E_{min}} & E(\mathbf{X}_i) \leq E_{ave} \\ w_{min} & E(\mathbf{X}_i) > E_{ave} \end{cases} \tag{26}$$

where w_{max} and w_{min} are the maximum and minimum values of the inertia weight, respectively, E_{min} is the minimum fitness of all particles in each iteration, E_{ave} is the average fitness of all particles in each iteration, and $E(\mathbf{X}_i)$ is the fitness value of the current particle in each iteration.

According to the description above, the actual kinematic parameter identification procedure of the parallel robot based on the improved PSO algorithm is outlined in Figure 2.

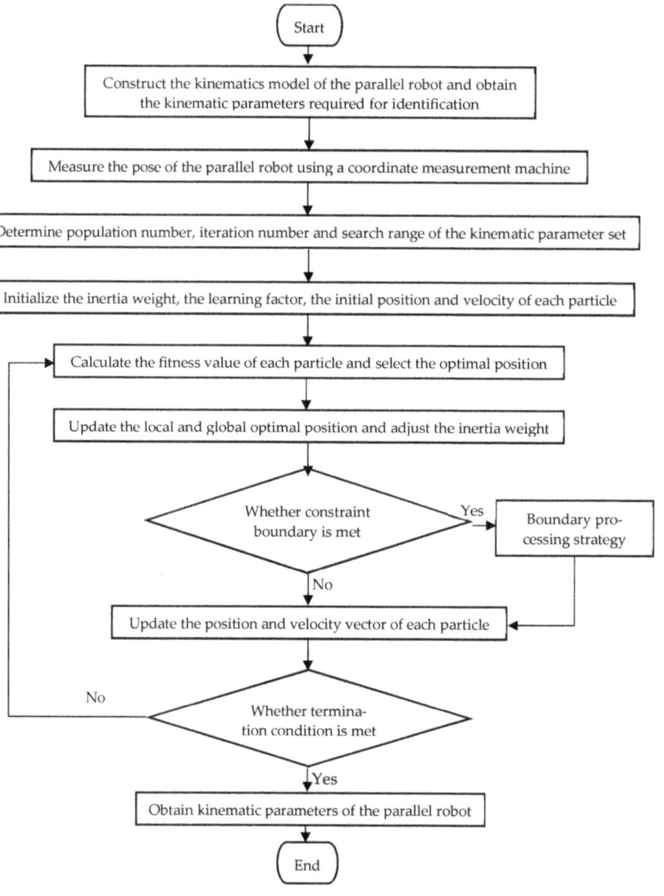

Figure 2. Flow chart of the kinematic parameter identification method of the parallel robot.

4. Simulations and Experiments

4.1. Identification Simulations

In order to verify the effectiveness of the proposed kinematic parameter identification method of the parallel robot, identification simulations of the parallel robot have been carried out. The original nominal kinematic parameters of the parallel robot are given in Table 1. According to the objective function L of the parallel robot, there are forty-two kinematic parameters to be identified in the identification simulations. Therefore, three actual kinematic parameter sets with the different normal distribution errors are listed in Tables 2–4. The five Gaussian measurement noises with variances of 0.000001 mm, 0.00001 mm, 0.0001 mm, 0.001 mm, and 0.01 mm were added to the actual poses of the parallel robots. Meanwhile, the influence of different numbers and configurations of the measurement poses on the identification accuracy was considered. The number of the measurement poses was set as 7, 14, 21, and 42, respectively. The configuration of the measurement poses was implemented by using the random selection method (Method I), the boundary exploration method [57] (Method II), and the optimal experiment design method [58] (Method III). Thus, the twelve pose sets were obtained and recorded in order, as shown in Table 5.

Table 1. The original nominal kinematic parameters of the parallel robot.

i	a_i/mm	b_i/mm	l_{0i}/mm
1	557.58, 52.02, 0	738.55, 945.81, 0	1830
2	−233.74, 508.89, 0	449.82, 111.50, 0	1830
3	−323.84, 456.87, 0	−1188.40, 166.70, 0	1830
4	−323.84, −456.87, 0	−1188.40, −166.70, 0	1830
5	−233.74, −508.89, 0	449.82, −1112.50, 0	1830
6	557.58, −52.02, 0	738.55, −945.81, 0	1830

Table 2. The assumptive actual kinematic parameters with the normal distribution error N (0, 10^{-3}).

i	a_i/mm	b_i/mm	l_{0i}/mm
1	557.5788, 51.9946, 0.0028	738.5480, 945.8080, −0.0031	1829.9969
2	−233.7381, 508.8839, −0.0083	449.8254, 1112.5011, −0.0112	1830.0046
3	−323.8524, 456.8764, 0.0115	−1188.4016, 166.7066, −0.0255	1830.0125
4	−323.8502, −456.8712, −0.0078	−1188.4114, −166.6968, −0.0057	1829.9837
5	−233.7494, −508.8928, −0.0085	449.8112, −1112.4949, −0.0073	1830.0019
6	557.5761, −51.9997, 0.0013	738.5440, −945.8007, 0.0145	1830.0017

Table 3. The assumptive actual kinematic parameters with the normal distribution error N (0, 10^{-2}).

i	a_i/mm	b_i/mm	l_{0i}/mm
1	557.8647, 51.9814, 0.0150	738.5183, 945.7937, 0.0819	1830.0082
2	−233.5444, 508.9095, −0.2333	449.7606, 1112.6096, −0.0195	1829.8084
3	−323.9119, 456.6571, 0.2699	−1188.2549, 166.8153, 0.0393	1830.0799
4	−324.0315, −456.9286, 0.0059	−1188.4592, −166.9436, 0.0188	1829.9816
5	−233.7380, −508.9476, −0.0801	449.7894, −1112.3236, 0.0136	1829.9522
6	557.4540, −51.9589, −0.0060	738.5899, −945.9340, 0.0500	1829.5340

Table 4. The assumptive actual kinematic parameters with the normal distribution error N (0, 1).

i	a_i/mm	b_i/mm	l_{0i}/mm
1	557.5648, 51.6396, −0.4139	738.7311, 947.0077, −1.0322	1828.9041
2	−235.3959, 510.2699, 0.6826	449.2160, 1112.0960, 0.4898	1830.6860
3	−324.8924, 459.1436, 0.9307	−1189.4993, 167.7338, 0.8544	1830.4674
4	−323.7261, −457.4479, 1.8226	−1188.0030, −167.2018, −1.1781	1832.1990
5	−234.2453, −508.6818, 0.1191	451.1179, −1111.8838, −1.4691	1830.2927
6	557.5536, −51.8896, 0.4607	737.1744, −946.7343, −1.8102	1829.5358

Table 5. The pose set number.

Method / Quantity	Method I	Method II	Method III
7	Pose set 1	Pose set 2	Pose set 3
14	Pose set 4	Pose set 5	Pose set 6
21	Pose set 7	Pose set 8	Pose set 9
42	Pose set 10	Pose set 11	Pose set 12

In the identification simulations, the objective function L was solved by the improved PSO algorithm, and the actual kinematic parameters were obtained. The identification error is defined as the root mean square of the difference between the actual and identified kinematic parameters to evaluate the global optimization ability and the identification precision of the proposed kinematic parameter identification method. The influence of different pose sets, measurement noises, and kinematic parameter sets on the identification errors is shown in Figures 3–5. For the same measurement noise and kinematic parameter set, the identification errors are greatly reduced with the increasing number of the measurement poses. In addition, it can be seen that although the pose number in pose set 10 is greater than that in pose set 9, the identification error of pose set 9 is less than that of pose set 10. The identification error of the pose sets based on the D-optimal method is more reduced while the number of the measurement poses is constant.

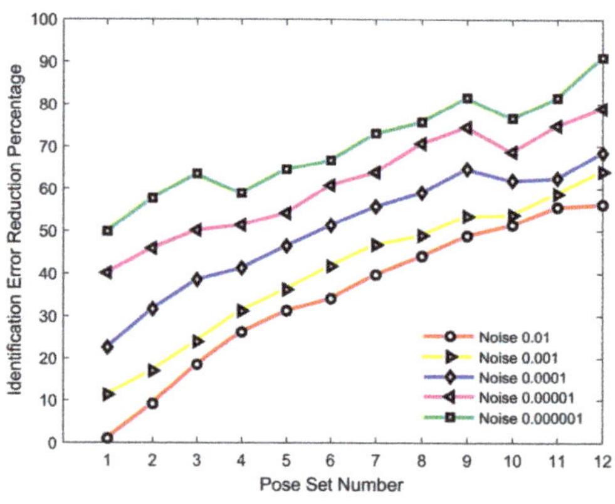

Figure 3. Correlation between identification error and pose set for parameter set 1.

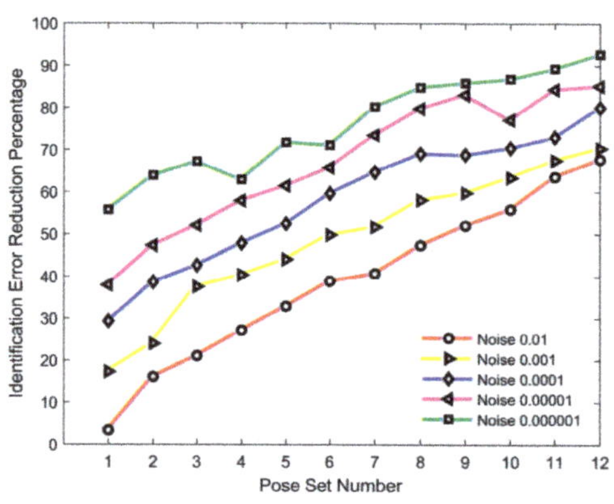

Figure 4. Correlation between identification error and pose set for parameter set 2.

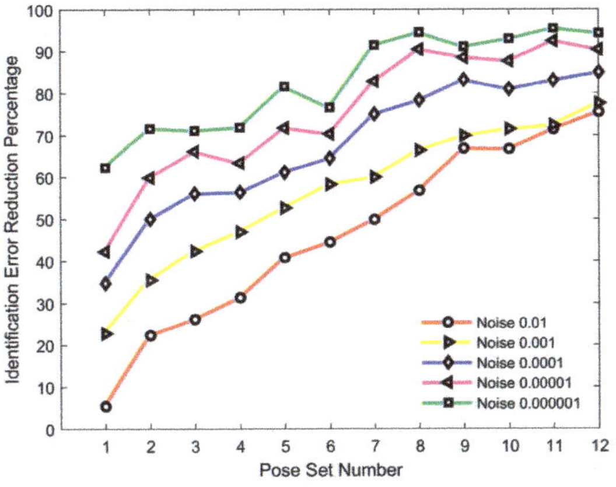

Figure 5. Correlation between identification error and pose set for parameter set 3.

The reduction in the identification error is not the ultimate goal of the kinematic parameter identification method of the parallel robot. The aim of the proposed kinematic parameter identification method is to enhance the positioning and orientation accuracy of the parallel robot. In order to describe the influence of the identification error on enhancing the positioning and orientation accuracy of the parallel robot, two accuracy indices (RMSPE and RMSOE) are introduced and expressed as follows:

$$\begin{aligned} RMSPE &= \sqrt{\frac{1}{n}\sum_{i=1}^{M}\left(e_{i(1)}^2 + e_{i(2)}^2 + e_{i(3)}^2\right)} \\ RMSOE &= \sqrt{\frac{1}{n}\sum_{i=1}^{M}\left(e_{i(4)}^2 + e_{i(5)}^2 + e_{i(6)}^2\right)} \end{aligned} \tag{27}$$

The relationship between positioning error reduction and identification error is shown in Figure 6, and the relationship between orientation error reduction and identification error is shown in Figure 7. Note that the positioning error of the parallel robot decreases as the percentage of identification error reduction increases. Similarly, the decrease of orientation error depends on the percentage of identification error reduction in a linear way. It is indicated that the proposed kinematic parameter identification method can improve the positioning and orientation accuracy of the parallel robot.

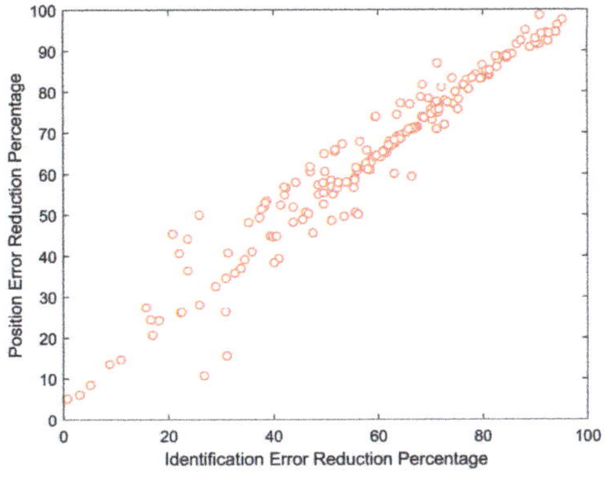

Figure 6. Correlation between identification error and position error.

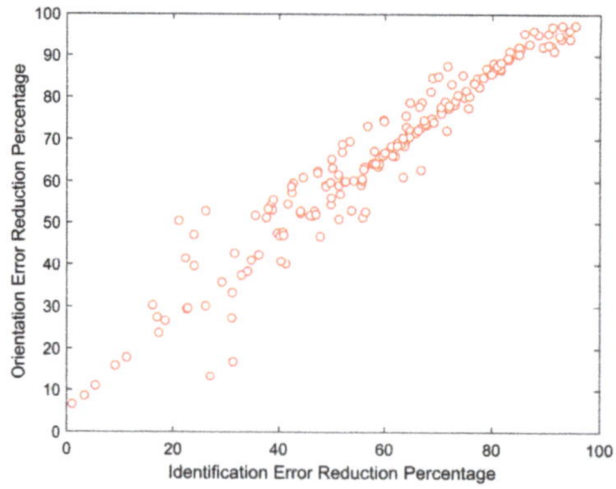

Figure 7. Correlation between identification error and orientation error.

The improvements of positioning and orientation accuracy affected by different measurement noise are shown in Figures 8–13. The positioning and orientation accuracy of the parallel robot can be improved with the reduction in measurement noise. Regardless of the magnitude of the initial kinematic parameter errors, the proposed kinematic parameter identification method can be perfectly performed while the measurement noise is small enough. However, the ultimate positioning and orientation accuracy of the parallel robot are not determined by the measurement noise alone. For different kinematic parameter sets, the curves in Figures 8–13 are significantly not identical. In addition, the larger the initial kinematic parameter errors of the parallel robot, the larger the identified kinematic parameter errors obtained by the proposed kinematic parameter identification method while the measurement noise is held constant. The larger the initial kinematic parameter errors of the parallel robot are, the more the ultimate positioning and orientation errors will be reduced while the same measurement pose number and configuration selection method are used in the identification simulation. The positioning and orientation accuracy of the parallel robot can be enhanced by the above three measurement pose configuration selection methods for any kinematic parameter error set. Among the three methods, the optimal experiment design method is the best to improve the positioning and orientation accuracy of the parallel robot.

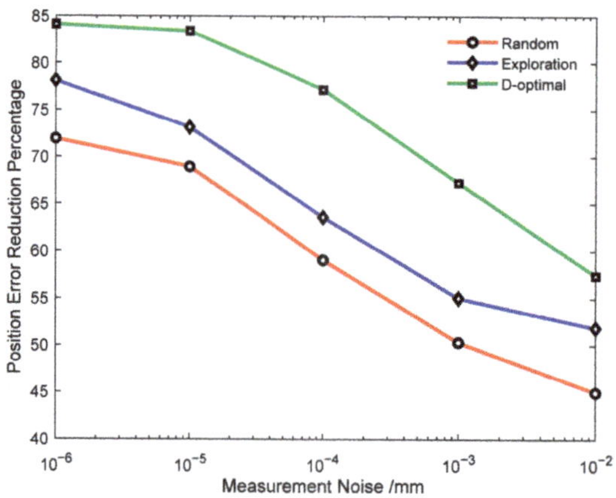

Figure 8. Correlation between position error and measurement noise for parameter error set 1.

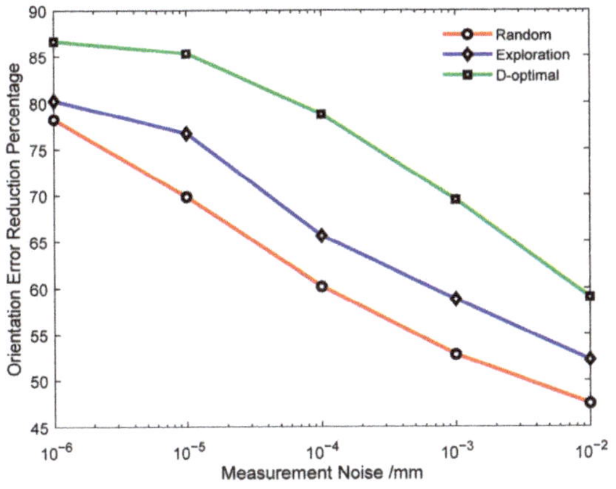

Figure 9. Correlation between orientation error and measurement noise for parameter error set 1.

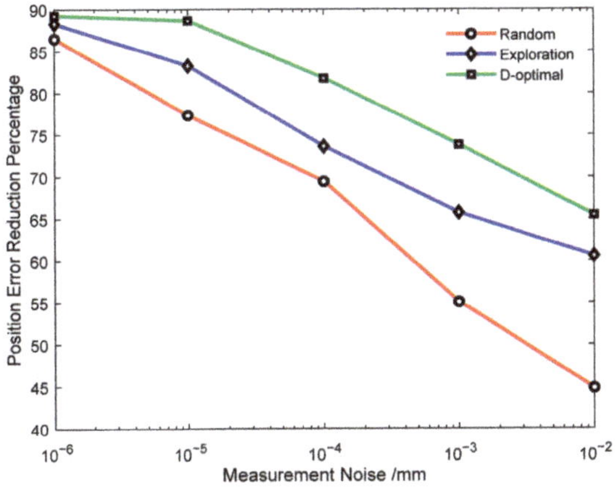

Figure 10. Correlation between position error and measurement noise for parameter error set 2.

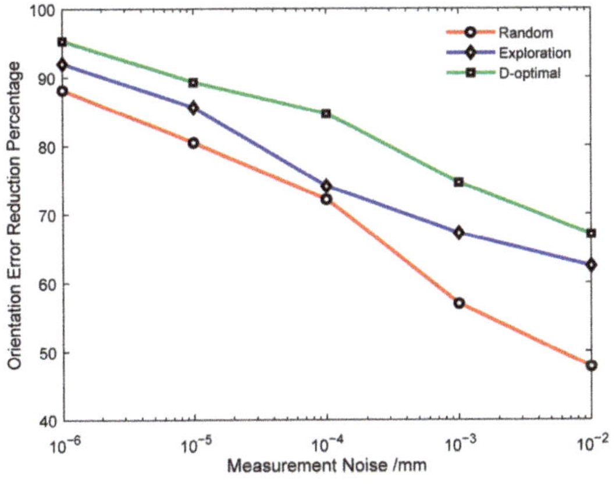

Figure 11. Correlation between orientation error and measurement noise for parameter error set 2.

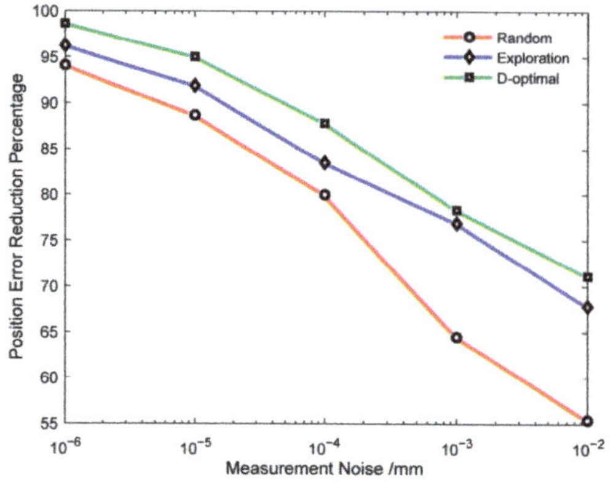

Figure 12. Correlation between position error and measurement noise for parameter error set 3.

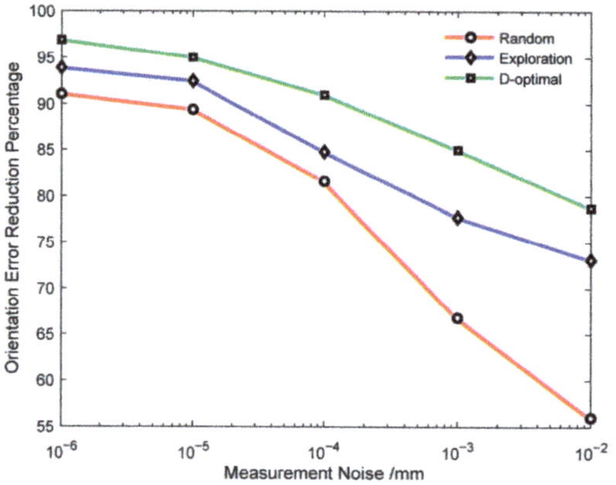

Figure 13. Correlation between orientation error and measurement noise for parameter error set 3.

4.2. Calibration Experiments

According to the identification simulation results, 42 different poses obtained using the D-optimal method were measured by a coordinate measuring machine, and the corresponding stroke lengths of the hydraulic cylinders were obtained from the controller of the parallel robot. The experimental system built in Harbin Institute of Technology involving a parallel robot, a coordinate measuring machine, and three feature spheres for pose measurement is shown in Figure 14 and described in [59]. The kinematic parameter identification of the parallel robot was implemented by using the improved PSO algorithm and nonlinear least squares (NLS) algorithm. After the actual kinematic parameters of the parallel robot were identified, 100 poses were randomly selected in the workspace and independently used to demonstrate the positioning and orientation accuracy of the parallel robot after kinematic calibration. The 100 poses were remeasured, and the corresponding positioning and orientation errors were recalculated. The absolute positioning and orientation errors of the parallel robot before and after kinematic calibration are shown in Figures 15 and 16. The maximum value (Max), average value (Mean), standard deviation value (STD), and root mean square value (RMS) of the positioning and orientation errors of the parallel robot before and after kinematic calibration of the 100 poses are shown in Table 6.

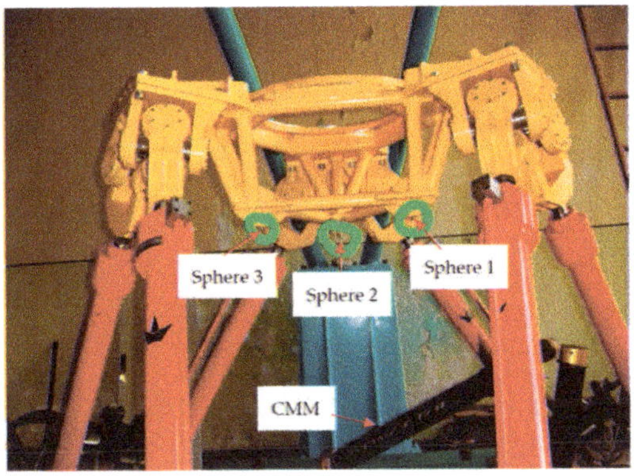

Figure 14. The experimental system built in HIT.

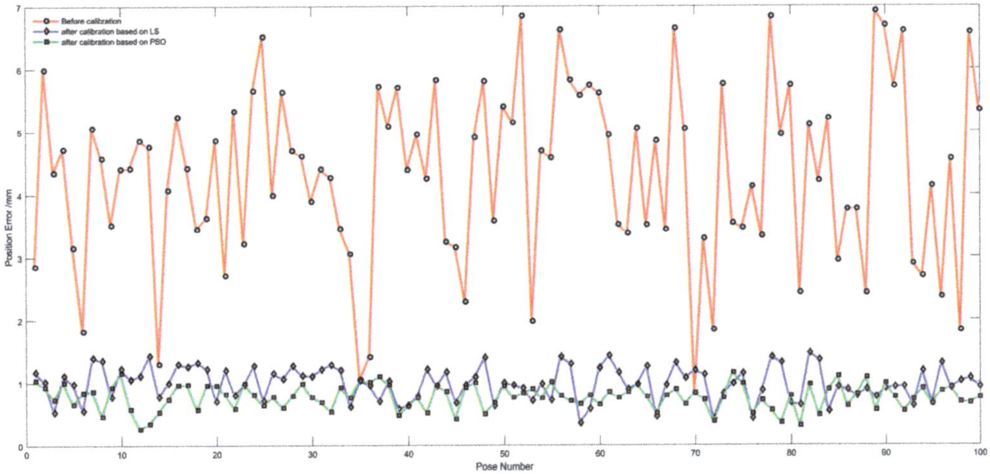

Figure 15. Absolute position errors before and after kinematic calibration.

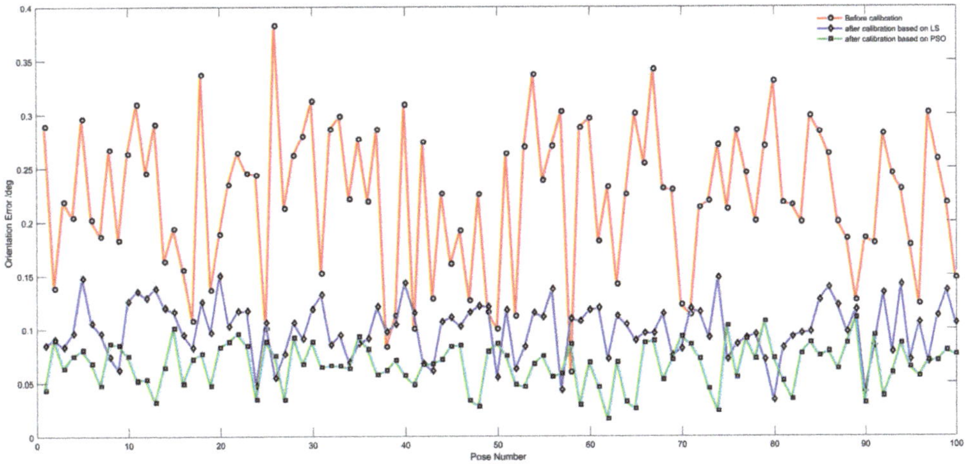

Figure 16. Absolute orientation errors before and after kinematic calibration.

Table 6. Pose error indices before and after kinematic calibration.

	Before Kinematic Calibration		After Kinematic Calibration			
			NLS Algorithm		PSO Algorithm	
	Position Error (mm)	Orientation Error (deg)	Position Error (mm)	Orientation Error (deg)	Position Error (mm)	Orientation Error (deg)
Max	6.9214	0.3825	1.4676	0.1502	1.1581	0.1105
Mean	4.3268	0.2221	0.9895	0.1003	0.7692	0.0674
Std	1.4179	0.0688	0.2705	0.0253	0.1997	0.0206
RMS	4.5510	0.2324	1.0255	0.1034	0.7944	0.0704

The maximum positioning error after kinematic calibration based on the NLS algorithm is 1.4676 mm, and the maximum based on the improved PSO algorithm is 1.1581 mm. The maximums of positioning error are reduced by 78.79% and 83.26%, respectively. The maximums of orientation error are 0.1502 and 0.1105 deg, respectively. Accordingly, the maximums of orientation error are reduced by 60.73% and 71.11%. The mean of positioning error decreased from 4.3269 to 0.9895 and 0.7692 mm, respectively. The mean of orientation error is decreased from 0.2221 to 0.1003 and 0.0674 deg, respectively. Although the two kinematic parameter identification methods can enhance the position and orientation accuracy of the parallel robot, the kinematic parameter identification method based on the improved PSO algorithm is more precise and effective than that based on the NLS algorithm.

5. Conclusions

The positioning and orientation accuracy requirements of the parallel robot in a docking motion simulation system are much higher than those of industrial motion simulators. In order to improve positioning and orientation accuracy of the parallel robot, a new kinematic parameter identification based on an improved PSO algorithm was proposed to identify the actual kinematic parameters of the parallel robot. The effectiveness and preponderance of the proposed kinematic parameter identification method were demonstrated by identification simulations and calibration experiments.

According to the identification simulation results, all twelve pose sets with different measurement noise levels are effective to estimate the different kinematic parameter sets. It is shown that the pose set 12 is the optimal pose set selection for the kinematic parameter identification of the parallel robot. From the calibration experiment results, the proposed kinematic parameter identification method increases the mean of position accuracy by 22.26% and orientation accuracy by 32.80% compared with the nonlinear least squares method, respectively. In conclusion, the kinematic parameter identification method based on the improved PSO algorithm can be used to enhance positioning and orientation accuracy of the parallel robot in a docking motion simulation system.

In future research, the influence of Hooke's joint and hydraulic cylinder compliance on the positioning and orientation errors will also be considered in error modeling, and it is necessary to create a practical kinematic calibration framework for improving the positioning and orientation accuracy of the parallel robot under loaded conditions.

Funding: This research received no external funding.

Data Availability Statement: The data presented in this study are available upon request from the author.

Conflicts of Interest: The author declares no conflicts of interest.

References

1. Huang, Q.T.; Jiang, H.Z.; Zhang, S.Y.; Han, J.W. Spacecraft docking simulation using hardware-in-the-loop simulator with Stewart platform. *Chin. J. Mech. Eng.* **2005**, *18*, 415–418. [CrossRef]
2. Wu, D.S.; Gu, H.B. Adaptive fuzzy backstepping control of light flight simulator motion platform. *Chin. J. Sci. Instrum.* **2008**, *29*, 1464–1473.

3. Zhang, C.; Zhang, L.Y. Kinematics analysis and workspace investigation of a novel 2-DOF parallel manipulator applied in vehicle driving simulator. *Robot. Comput.-Integr. Manuf.* **2013**, *29*, 113–120. [CrossRef]
4. Wang, Q.; Qi, C.K.; Gao, F.; Zhao, X.C.; Ren, A.Y.; Hu, Y. Normal contact stiffness identification-based force compensation for a hardware-in-the-loop docking simulator. *Adv. Robot.* **2018**, *5*, 266–282. [CrossRef]
5. Harandi, M.R.J.; Khalilpour, S.A.; Taghirad, H.D. Adaptive energy shaping control of a 3-DOF underactuated cable-driven parallel robot. *IEEE Trans. Ind. Inform.* **2023**, *19*, 7552–7560. [CrossRef]
6. Bamberger, H.; Shoham, M. A novel six degrees-of-freedom parallel robot for MEMS fabrication. *IEEE Trans. Robot.* **2007**, *23*, 189–195. [CrossRef]
7. Li, C.Y.; Wu, H.P.; Eskelinen, H.; Ji, H.B. Mobile parallel robot machine for heavy-duty double-walled vacuum vessel assembly. *Ind. Robot-Int. J. Robot. Res. Appl.* **2021**, *48*, 523–531. [CrossRef]
8. Xie, Z.H.; Xie, F.G.; Liu, X.J.; Wang, J.S.; Mei, B. Tracking error prediction informed motion control of a parallel machine tool for high-performance machining. *Int. J. Mach. Tools Manuf.* **2021**, *164*, 103714. [CrossRef]
9. Tanaka, H.; Morimoto, Y.; Hayashi, A.; Yamaoka, H. Posture evaluation based on forward kinematics and inverse kinematics of parallel link type machine tool. *Int. J. Autom. Technol.* **2022**, *16*, 497–506. [CrossRef]
10. Wang, S.L.; Zou, H.X.; Wang, S.B. A Tabu-GA-based parallel machine scheduling with restrained tool resources. *Proc. Inst. Mech. Eng. Part B-J. Eng. Manuf.* **2022**, *236*, 39–50. [CrossRef]
11. Meng, C.R.; Wang, N.; Li, Q.; Li, X.N.; Liu, Q.Y.; Gao, P. A guiding robot for minimally invasive spinal surgery based on parallel manipulator. *Adv. Mater. Res.* **2014**, *902*, 280–285. [CrossRef]
12. Zhang, N.B.; Huang, P.C.; Li, Q.C. Modeling, design and experiment of a remote-center-of motion parallel manipulator for needle insertion. *Robot. Comput.-Integr. Manuf.* **2018**, *50*, 193–202. [CrossRef]
13. Saracino, A.; Oude-Vrielink, T.J.C.; Menciassi, A.; Sinibaldi, E.; Mylonas, G.P. Haptic intracorporeal palpation using a cable-driven parallel robot: A user study. *IEEE Trans. Biomed. Eng.* **2020**, *67*, 3453–3463. [CrossRef] [PubMed]
14. Ulinici, L.; Vaida, C.; Gherman, B.; Antal, T.; Tucan, P.; Pisla, D. Kinematics and workspace simulation of a new parallel robot for sils. *Acta Techanica Napoc. Ser.-Appl. Math. Mech. Eng.* **2022**, *65*, 505–514.
15. Song, Y.M.; Jia, D.K.; He, Z.Y.; Lian, B.B.; Sun, T.; Yao, Q. Safety motion control for a parallel fracture surgical robot. *J. Tianjin Univ.* **2023**, *56*, 221–231.
16. Brahmia, A.; Kelaiaia, R.; Company, O.; Chemori, A. Kinematic sensitivity of manipulators using a novel dimensionless index. *Robot. Auton. Syst.* **2022**, *150*, 104021. [CrossRef]
17. Kelaiaia, R.; Chemori, A.; Brahmia, A.; Kerboua, A. Optimal dimensional design of parallel manipulators with an illustrative case study: A review. *Mech. Mach. Theory* **2023**, *188*, 105390. [CrossRef]
18. Brahmia, A.; Kerboua, A.; Kelaiaia, R.; Latreche, A. Tolerance synthesis of Delta-like parallel robots using a nonlinear optimization method. *Appl. Sci.* **2023**, *13*, 10703. [CrossRef]
19. Zhuang, H.Q.; Roth, Z.S. A linear solution to the kinematic parameter identification of robot manipulators. *IEEE Trans. Robot. Autom.* **1993**, *9*, 174–185. [CrossRef]
20. He, R.B.; Zhao, Y.J.; Yang, S.N.; Yang, S.Z. Kinematic-parameter identification for serial-robot calibration based on POE formula. *IEEE Trans. Robot.* **2010**, *26*, 411–423.
21. Everett, L.J.; Hsu, T.W. The theory of kinematic parameter identification for industrial robots. *J. Dyn. Syst. Meas. Control* **1988**, *110*, 96–100. [CrossRef]
22. Huang, T.; Chetwynd, D.G.; Whitehouse, D.J.; Wang, J.S. A general and novel approach for parameter identification of 6-DOF parallel kinematic machines. *Mech. Mach. Theory* **2005**, *40*, 219–239. [CrossRef]
23. Verner, M.; Xi, F.F.; Mechefske, C. Optimal calibration of parallel kinematic machines. *J. Mech. Des.-Trans. ASME* **2005**, *127*, 62–69. [CrossRef]
24. Ni, Y.B.; Zhang, B.; Guo, W.X.; Shao, C.Y. Kinematic calibration of parallel manipulator with full-circle rotation. *Ind. Robot: Int. J.* **2016**, *43*, 296–307. [CrossRef]
25. Liu, Y.Z.; Wu, J.; Wang, L.P.; Wang, J.S.; Wang, D.; Yu, G. Kinematic calibration of a 3-DOF parallel tool head. *Ind. Robot: Int. J.* **2017**, *44*, 231–241. [CrossRef]
26. Fu, L.; Yang, M.; Liu, Z.H.; Tao, M.; Cai, C.G.; Huang, H. Stereo vision-based kinematic calibration method for the Stewart platforms. *Opt. Express* **2022**, *30*, 479597. [CrossRef] [PubMed]
27. He, Z.Y.; Song, Y.M.; Lian, B.B.; Sun, T. Kinematic calibration of a 6-DOF parallel manipulator with random and less measurements. *IEEE Trans. Instrum. Meas.* **2023**, *72*, 7500912. [CrossRef]
28. Li, J.; Yu, L.D.; Sun, J.Q.; Xia, H.J. A kinematic model for parallel-joint coordinate measuring machine. *J. Mech. Robot.-Trans. ASME* **2013**, *5*, 044501. [CrossRef]
29. Mao, C.T.; Chen, Z.W.; Li, S.; Zhang, X. Separable nonlinear least squares algorithm for robust kinematic calibration of serial robots. *J. Intell. Robot. Syst.* **2021**, *101*, 2. [CrossRef]
30. Nategh, M.J.; Agheli, M.M. A total solution to kinematic calibration of hexapod machine with a minimum number of measurement configurations and superior accuracies. *Int. J. Mach. Tools Manuf.* **2009**, *49*, 1155–1164. [CrossRef]
31. Majarena, A.C.; Santolaria, J.; Samper, D.; Aguilar, J.J. Analysis and evaluation of objective functions in kinematic calibration of parallel mechanism. *Int. J. Adv. Manuf. Technol.* **2013**, *66*, 751–761. [CrossRef]

32. Luo, J.B.; Chen, S.L.; Fan, X.N.; Xiong, C.Y.; Zheng, T.J.; Zhang, C.; Yang, G.L. Kinematic calibration of a 4PPa-2PaR parallel mechanism with subchains on limbs. *IEEE Trans. Instrum. Meas.* **2022**, *71*, 7502011. [CrossRef]
33. Nguyen, H.N.; Zhou, L.; Kang, H.J. A calibration method for enhancing robot accuracy through integration of an extended Kalman filter algorithm and an artificial neural network. *Neurocomputing* **2015**, *151*, 996–1005. [CrossRef]
34. Qi, J.D.; Chen, B.; Zhang, D.H. A calibration method for enhancing robot accuracy through integration of kinematic model and spatial interpolation algorithm. *J. Mech. Robot.-Trans. ASME* **2021**, *13*, 061013. [CrossRef]
35. Yin, F.W.; Wang, L.; Tian, W.J.; Zhang, X.P. Kinematic calibration of a 5-DOF hybrid machining robot using an extended Kalman filter method. *Precis. Eng.* **2023**, *79*, 86–93. [CrossRef]
36. Wang, Z.P.; Cao, B.S.; Xie, Z.W.; Ma, B.Y.; Sun, K.; Liu, Y. Kinematic calibration of a space manipulator based on visual measurement system with extended Kalman filter. *Machines* **2023**, *11*, 409. [CrossRef]
37. Yang, W.Y.; Li, S.; Li, Z.B.; Luo, X. Highly accurate manipulator calibration via extended Kalman filter-incorporated residual neural network. *IEEE Trans. Ind. Inform.* **2023**, *19*, 10831–10841. [CrossRef]
38. Du, G.L.; Zhang, P. Online serial manipulator calibration based on multisensory process via extended Kalman and particle filters. *IEEE Trans. Ind. Electron.* **2014**, *61*, 6852–6859.
39. Jiang, Z.H.; Zhou, W.G.; Li, H.; Mo, Y.; Ni, W.C.; Huang, Q. A new kind of accurate calibration method for robotic kinematic parameters based on the extended Kalman and particle filter algorithm. *IEEE Trans. Ind. Electron.* **2018**, *65*, 3337–3345. [CrossRef]
40. Du, G.L.; Liang, Y.H.; Li, C.Q.; Liu, P.X.; Li, D. Online robot kinematic calibration using hybrid filter with multiple sensors. *IEEE Trans. Instrum. Meas.* **2020**, *69*, 7092–7107. [CrossRef]
41. Gao, G.B.; Li, Y.; Liu, F.; Han, S.C. Kinematic calibration of industrial robots based on distance information using a hybrid identification method. *Complexity* **2021**, *2021*, 8874226. [CrossRef]
42. Feng, C.S.; Cong, S.; Shan, H.; Shang, W.W.; Zhang, Y.X. Self-calibration for kinematic parameters of a redundant planar two-degree-of-freedom parallel manipulator using evolutionary algorithm. *Eng. Optim.* **2009**, *41*, 385–398. [CrossRef]
43. Sun, T.; Lian, B.B.; Zhang, J.T.; Song, Y.M. Kinematic calibration of 2-DoF over-constrained parallel mechanism using real inverse kinematics. *IEEE Access* **2018**, *6*, 67752–67761. [CrossRef]
44. Fei, Z.G.; Xu, X.J.; Xiao, Y.Q.; Meier, N.; Georgiadis, A. Kinematic self-calibration of non-contact five-axis measuring machine using improved genetic algorithm. *Meas. Sci. Technol.* **2016**, *27*, 025903. [CrossRef]
45. Fan, C.; Zhao, G.L.; Zhao, J.; Zhang, L.; Sun, L.N. Calibration of a parallel mechanism in a serial-parallel polishing machine tool based on genetic algorithm. *Int. J. Adv. Manuf. Technol.* **2015**, *81*, 27–37. [CrossRef]
46. Huang, L.E.; Chen, M.F.; Zheng, S.G.; He, C.Y.; Zhu, E.X.; Zhang, Y.X. Research on the kinematic calibration of the 3-PTT parallel mechanism. *J. Mech. Sci. Technol.* **2023**, *37*, 4311–4325. [CrossRef]
47. Gao, G.B.; Liu, F.; San, H.J.; Wu, X.; Wang, W. Hybrid optimal kinematic parameter identification for an industrial robot based on BPNN-PSO. *Complexity* **2018**, *2018*, 4258676. [CrossRef]
48. Wang, W.D.; Song, H.J.; Yan, Z.Y.; Sun, L.N.; Du, Z.J. A universal index and an improved PSO algorithm for optimal pose selection in kinematic calibration of a novel surgical robot. *Robot. Comput.-Integr. Manuf.* **2017**, *50*, 90–101. [CrossRef]
49. Li, H.; Hu, X.B.; Zhang, X.J.; Wei, S.Y. Kinematic parameters calibration of industrial robot based on RWS-PSO algorithm. *J. Mech. Eng. Sci.* **2023**, *237*, 3210–3220. [CrossRef]
50. Chen, X.Z.; Zhan, Q. The kinematic calibration of an Industrial robot with an improved beetle swarm optimization algorithm. *IEEE Robot. Autom. Lett.* **2022**, *7*, 4694–4701. [CrossRef]
51. Zhong, H.R.; Hu, C.H.; Li, X.Y.; Gao, L.; Zeng, B.; Dong, H.Z. Kinematic calibration method for a two-segment hydraulic leg based on an improved whale swarm algorithm. *Robot. Comput.-Integr. Manuf.* **2019**, *59*, 361–372. [CrossRef]
52. Gong, C.K.; Sun, Y.H.; Yuan, L.P. Study on application of hybrid optimization algorithm in parameters optimization of manipulator. *J. Syst. Simul.* **2018**, *30*, 105–113.
53. Fang, L.J.; Dang, P.F. Kinematic calibration method of robots based on quantum-behaved particle swarm optimization. *J. Mech. Eng.* **2016**, *7*, 23–30. [CrossRef]
54. Yang, C.F.; Zheng, S.T.; Jin, J.; Zhu, S.B.; Han, J.W. Forward kinematics analysis of parallel manipulator using modified global newton-Raphson method. *J. Cent. South Univ. Technol.* **2010**, *17*, 1264–1270. [CrossRef]
55. Kennedy, J.; Eberhart, R. Particle Swarm Optimization. In Proceedings of the 1995 IEEE International Conference on Neural Networks 1995, Perth, WA, Australia, 27 November–1 December 1995; pp. 1942–1948.
56. Zhou, B.; Li, S.P.; Zi, B.; Chen, B.; Zhu, W.D. Multi-objective optimal design of a cable-driven parallel robot based on an adaptive adjustment inertia weight particle swarm optimization algorithm. *J. Mech. Des.* **2023**, *145*, 083301. [CrossRef]
57. Patel, A.J.; Ehmann, K.F. Calibration of a hexapod machine tool using a redundant leg. *Int. J. Mach. Tools Manuf.* **2000**, *40*, 489–512. [CrossRef]
58. Kamli, K.; Bonev, I.A. Optimal experiment design for elasto-geometrical calibration of industrial robots. *IEEE-ASME Trans. Mechatron.* **2019**, *24*, 2733–2744. [CrossRef]
59. Yu, D.Y. Kinematic calibration of parallel robots based on L-Infinity parameter estimation. *Machines* **2022**, *10*, 436. [CrossRef]

MDPI AG
Grosspeteranlage 5
4052 Basel
Switzerland
Tel.: +41 61 683 77 34

Applied Sciences Editorial Office
E-mail: applsci@mdpi.com
www.mdpi.com/journal/applsci

www.ingramcontent.com/pod-product-compliance
Lightning Source LLC
Chambersburg PA
CBHW061609210926
51850CB00009B/2263
9783725874347